Research Methodology

Communications in Cybernetics, Systems Science and Engineering

ISSN: 2164-9693

Book Series Editor:

Jeffrey Yi-Lin Forrest

International Institute for General Systems Studies, Grove City, USA
Slippery Rock University, Slippery Rock, USA

Volume 3

Research Methodology

From Philosophy of Science to Research Design

Alexander M. Novikov
Research Center of the Theory of Continuous Education,
Russian Academy of Education, Moscow, Russian Federation

Dmitry A. Novikov
Trapeznikov Institute of Control Sciences, Russian Academy of Sciences,
Moscow, Russian Federation

CRC Press
Taylor & Francis Group
Boca Raton London New York

CRC Press is an imprint of the
Taylor & Francis Group, an **informa** business

A BALKEMA BOOK

CRC Press
Taylor & Francis Group
6000 Broken Sound Parkway NW, Suite 300
Boca Raton, FL 33487-2742

First issued in paperback 2019

© 2013 by Taylor & Francis Group, LLC
CRC Press is an imprint of Taylor & Francis Group, an Informa business

Typeset by MPS Limited, Chennai, India

ISBN-13: 978-1-138-00030-8 (hbk)
ISBN-13: 978-0-367-38012-0 (pbk)

Library of Congress Cataloging-in-Publication Data

Visit the Taylor & Francis Web site at
http://www.taylorandfrancis.com

and the CRC Press Web site at
http://www.crcpress.com

Table of contents

Editorial board

About the authors

Alexander M. Novikov was born in 1941. Honored Scientist of the Russian Federation, Dr. Sci. (Pedagogics), Prof., academic of the Russian Academy of Education, foreign member of the Ukrainian Academy of Pedagogical Sciences, member of the Union of Journalists, laureate of the National Prize of the Russian Federation.

At present, he is head of the Research Center of the theory of continuous education of the Russian Academy of Education. He has authored over 300 scientific publications on: methodology and the theory of pedagogics, the theory and methods of labour education and professional education, psychology and physiology of labour. Scientific adviser of 10 Doctors of Science and 32 Candidates of Science. e-mail: amn@anovikov.ru, www.anovikov.ru.

Dmitry A. Novikov was born in 1970. Dr. Sci. (Eng.), Prof., corresponding member of the Russian Academy of Sciences. At present, he is Deputy Director of the Trapeznikov Institute of Control Sciences of the Russian Academy of Sciences, and Head of the Control Sciences Department of the Moscow Institute of Physics and Technology (MIPT).

He has authored over 400 scientific publications on the theory of control in interdisciplinary systems, including research works on: methodology, system analysis, game theory, decision-making and control mechanisms in social and economic systems. Scientific adviser of 6 Doctors of Science and 24 Candidates of Science. e-mail: novikov@ipu.ru, www.mtas.ru.

Introduction

Methodology is the theory of organization of an activity[1]. Such definition uniquely determinates the subject of methodology, which is organization of an activity. Within the framework of this unified approach, proposed and developed in [29], the methodologies of scientific activity, practical activity, educational activity, art activity, and play activity have been described to date.

Not all activities require being organized with application of methodology. A human activity can be divided into *imitative* activity and *productive* activity.

The imitative activity is a "cast," a copy of an activity of another person or a copy of one's own activity based on accumulated experience. For instance, the monotonous activity of a lathe operator in any machine workshop (at the level of mastered technologies) appears organized (self-organized) in principle. Evidently, such activity needs no application of methodology.

In contrast, the productive activity aims at obtaining an objectively new[2] or subjectively new result[3]. By definition, any scientific activity (being realized more or less competently) aims at an objectively new result. This is exactly the case of the productive activity which requires application of methodology.

Methodology being treated as the theory of organization of an activity, one should naturally consider the notion of an "organization." According to the definition provided by Merriam-Webster dictionary, an *organization* is:

1   The condition or manner of being organized;
2   The act or process of organizing or of being organized;
3   An administrative and functional structure (as a business or a political party); also, the personnel of such a structure. See Fig. I.1.

---

[1] There exist some narrower definitions of methodology. Notably, within the framework of the Cartesian paradigm, methodology is understood as the totality of methods to perform a certain activity. Sometimes, philosophers relate any general statements of a specific field of science or of a practical activity to the scope of methodology.

[2] A kind of activity intended for obtaining of an objectively new result is called creation.

[3] The so-called "arranging" activity is an activity which represents a counterpart to the productive activity (in a certain sense). Whereas the productive activity often breaks the former order and old stereotypes, the arranging activity aims at the order recovery (this is clear from its name). It consists in establishing some norms of activity implemented, in particular, in the form of standards, laws, orders, etc.

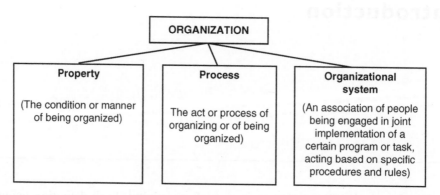

Figure I.I Definition of an organization.

Thus, we adopt mainly the first and second meaning of the notion of an organization; i.e., we consider it as both the *property* of being organized (the first meaning) and the *process* of organizing including the result of this process (the second meaning). The third meaning (an organizational *system*), will be also involved (to a smaller extent) in the description of collective scientific activity and management of scientific projects.

Let us outline the structure and logic of this book.

Methodology considers organization of an *activity*: an activity is the active behaviour of a human being. Organizing an activity means arranging it as an integral system with clearly defined characteristics, a logical structure and the accompanying process of its realization, the *temporal structure*. The corresponding reasoning lies in a pair of dialectic categories "historical (temporal)" and "logical."

The logical structure includes the following components of activity: subject, object, item, forms, means, methods, and result.

The following *characteristics of activity*: are external with respect to this structure: features, principles, conditions, and norms.

Various kinds of activity organization culture are historically established, see Chapter 1. Nowadays, we apply the project-technological kind – the productive activity of a human being (or an organization) is decomposed into separate completed cycles called *projects*[4].

The process of activity implementation is considered within the framework of a project realized in a time sequence by phases, stages and steps. Furthermore, this sequence is common for all kinds of activity. The completeness of an activity cycle (a project) is defined by the following three phases:

– *design phase*, which yields the model of a created system (a scientific hypothesis as the model of a created system of a new scientific knowledge) and the plan of its implementation;

---

[4]Today there exist two common definitions of a project. The first one implies that a project is the normative model of a certain system. The second definition states that a project is the purposeful creation or modification of a certain system, having a specific organization under constraints imposed on available time and resources. In this book we employ the second definition (see below).

- *technological phase*, which yields implementation of the system, i.e., verification of the hypothesis;
- *reflexive phase*, which yields an estimate of the constructed system of a new scientific knowledge and indicates the necessity of its further correction or "launching" of a new project (i.e., generating and verifying a new hypothesis).

Therefore, it is possible to suggest the following "**scheme of the methodology of scientific research**":

1 The characteristics of scientific activity:

- features,
- principles,
- conditions,
- norms of scientific activity;

2 The logical structure of scientific activity:

- subject,
- object,
- topic,
- forms,
- means,
- methods,
- result of scientific activity;

3 The temporal structure of scientific activity:

- phases,
- stages,
- steps of scientific activity.

The methodology of scientific research occupies an "intermediate" position (serves as a "bridge") in the following hierarchy:

- the philosophy of science;
- the methodology of scientific research;
- research design;
- a research technique.

A *research technique* is a set of certain methods, tools, algorithms, etc. to perform a specific research [3, 7, 8, 23, 34, 36]. *Research design* is the process of choosing a research technique [6, 10, 11, 16]. *Research methodology* deals with general laws and principles of organizing the research activity – choosing an efficient (adequate, rational) research technique [12, 17, 19, 26]. Finally, the *philosophy of science* [5, 38, 41] corresponds to overall universal framework for any scientific activity.

The fundamental difference of this book (as opposed to numerous works devoted to research methodology) consists in the integrated approach to scientific research. Notably, the latter is considered as a scientific project throughout all the levels – from the philosophy of science to research design.

The book possesses the following structure. The basic principles of methodology are discussed in Chapter 1. Next, Chapter 2 focuses on the characteristics of scientific activity. Means and methods of scientific research are studied in Chapter 3. Organization of the research implementation process and its temporal structure are described in Chapter 4. Chapter 5 deals with organization of collective scientific research. The Conclusion serves for summarizing the presented material in the form of consolidated analytical tables. Finally, in the Appendix the authors share their opinions concerning the role of science in modern society.

Chapter 1

Foundations of research methodology

A *foundation* is a sufficient condition of something (e.g., one may consider foundations of objective reality, cognition, an idea or activity).

Recall that we understand methodology as the theory of organization of an activity. Then it seems possible to identify the following foundations of modern methodology (including research methodology):

1 The philosophical-psychological theory of activity [24, 39].
2 *Systems analysis* [2, 4, 28] and *systems engineering* [13, 40] (the theory focused on the system of research or design methods for complex systems, as well as methods to find, plan and implement changes for eliminating the existing problems.
3 The science of science (the theory of science). In the first place, methodology is related to *epistemology* (the theory of cognition) and *semiotics* (the theory of signs).
4 Ethics of an activity.
5 Aesthetics of an activity.

Chapter 1 has the following structure. In Section 1.1 we discuss the philosophical-psychological foundations and systems engineering foundations of methodology. Next, Section 1.2 deals with the epistemological foundations, whereas ethical and aesthetical foundations are described in Section 1.3.

1.1 PHILOSOPHICAL, PSYCHOLOGICAL AND SYSTEMATIC FOUNDATIONS

Since methodology is viewed as the theory of organization of an activity, we should start with the basic notions connected with an activity.

An activity is an active interaction of a human being with an external environment, where the former acts as a subject exerting a purposeful impact on an object to satisfy his/her own needs [24].

In philosophy, a *subject* is defined as a bearer of the object-oriented practical activity and cognition (an individual or a social group); as the source of active behavior directed towards an object. According to *dialectics*, a subject is remarkable for inherent self-consciousness (indeed, he/she has mastered the world of culture created by the humanity – the tools of the domain-practical activity, the forms of a language, logical categories, the norms of aesthetical or moral judgements, etc.). The active behavior of a subject forms a condition ensuring that a certain fragment of objective reality acts as an object given to the subject in the forms of his/her activity.

Meanwhile, philosophy determines an *object* as the entity opposing a subject in his/her object-oriented practical activity and cognition activity. An object appears non-identical with the objective reality, merely acting as its part which interacts with a subject.

Philosophy studies an activity as the comprehensive way of a human life; accordingly, a human being is defined as an active being. The human activity covers material-practical and intelligent (spiritual) operations, external and internal processes. The human activity lies equally in thinking and working, in cognition process and human behavior. Through activity a human being reveals his/her own (special) role in the world, asserting oneself as a social being.

Psychology considers an activity as an important component of psyche. For instance, S.L. Rubinstein believed that psychology should investigate not the activity of a subject as such, but "psyche exclusively" (as a matter of fact, by exploring its essential objective relations and mediations, including activity analysis, see [39]). On the other hand, A. Leont'ev adhered to the opinion that the activity must be the subject of psychology, so far as psyche is indissolubly connected with the moments of activity that generate and mediate it [24].

Actually, *systems analysis* and systems engineering occupy an interdisciplinary or overdisciplinary position and may be treated as applied dialectics. Within the framework of these approaches, the activity is a complex system intended for preparing, substantiating and implementing solutions to complex problems of different character (e.g., political, social, economic, technical problems, etc.).

By comparing the above conceptions of the three scientific disciplines (*viz.*, philosophy, psychology, and systems analysis or systems engineering), one would easily choose the general *structure of activity* (see Fig. 1.1). It will be extensively used in the sequel.

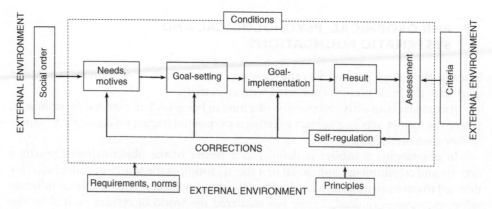

Figure 1.1 The structural components of an activity.

Let us consider the basic *structural components of any activity*.

Needs are defined as the requirement or lack of a certain entity being essential to sustain vital activity of an organism, an individual, a social group or society as a whole. Biological needs (in particular, human ones) are subject to metabolic conversion as a prerequisite for the existence of any living organism. The needs of social subjects, i.e., an individual, a social group and society as a whole, depend on the development level of a given society and on specific social conditions of their activity.

The needs are stated in concrete terms via *motives* that make a man or a social group act; in fact, activity is performed for the sake of motives. *Motivation* means the process of stimulating an individual or a social group to fulfill a specific activity, actions and steps. Motivation represents a complex process which requires analyzing certain alternatives, as well as choosing and making decisions.

Motives cause formation of a *goal* as a subjective image of the desired *result* of the expected activity or action. A *goal* has a special place in the activity structure. The key issue consists in the following. Who should assign goals? Suppose that goals are set externally (e.g., a lecturer gives tasks to a student, a manager assigns plans to a subordinate) or a certain person performs monotonous and routine work daily. In this case, activity possesses an imitative (executive), uncreative nature; consequently, no *goal-setting*[1] problem takes place (one should not choose a goal). Contrariwise, the productive activity (even a nonstandard activity and, *a fortiori*, an innovative or creative activity such as scientific research) is remarkable for that the subject directly determines the goal. As a result, the goal-setting process gets complicated; it includes specific stages and steps, as well as requires special methods and means. In terms of the project-technological type of organizational culture (see below) and in terms of systems analysis, the goal-setting process is defined as *design*. This notion will be used throughout the book.

The *goal-implementation* process is characterized by its content, forms, as well as by specific methods, means, and technologies.

A particular position within the activity structure is occupied by those components referred to as either self-regulation (in the case of an individual subject) or *control* (in the case of a collective subject).

Self-regulation is generally defined as reasonable functioning of living systems [2]. Psychical self-regulation is the regulation level for active behavior of such systems; psychical regulation expresses the specifics of psychical means of reality reflection and modeling (including reflexion of a subject). The notion of reflexion will be introduced later.

Self-regulation possesses the following structure: the goal of activity accepted by a subject – the model of relevant conditions of activity – the program of executive actions – the system of assessment criteria to be used for activity – information on the achieved results – appraisal of the achieved results in the sense of the assessment criteria – the decision regarding the necessity and character of corrections in activity.

Thus, self-regulation represents a closed control loop. This is an informational process whose medium include different forms of reality reflection.

---

[1]Note the processes of goal-setting and goal-implementation are described by internal conditions, forms, methods and means of their realization (see below).

Control is treated as an element, a function of organized systems of different nature (e.g., biological, social or technical ones), ensuring retention of their structure, maintenance of activity, and implementation of a program or a goal of activity [30]. Collective activity appears impossible without creating a definite order or division of labor, without establishing the place and functions of each man in a collective, being performed by means of control.

The notion of an *external environment* (see Fig. 1.1) turns out to be an essential category in system analysis. The external environment is defined as a set of those objects/subjects lying outside the system if, first, changes in their properties and/or behavior affect the system under consideration and, second, their properties and/or behavior change depending on behavior of the system [29].

In Fig. 1.1 we have separated the factors being set by an external environment (with respect to the given subject of activity). These are *criteria* used to assess the compliance of a result to a goal, *norms* (legal, ethical, hygienic, etc.) and *principles* of activity, widely adopted within a society or an organization. *Conditions of activity* (material and technical, financial, informational, etc.) are related to the external environment. At the same time, they can enter into the structure of activity (recall the feasibility of active influence of a subject on the conditions of his/her activity).

The following groups of **conditions** are invariant for any activity:

– motivational,
– personnel-related,
– material and technical,
– methodical,
– financial,
– organizational,
– regulatory and legal,
– informational.

However, in any concrete situation these groups may have specific features.

Thus, we have discussed primary characteristics of an activity and the corresponding structural components. Now, let us address the issues of methodology as the theory of organization of an activity.

In principle, human activity may be performed spontaneously, learning by one's own mistakes. **Methodology generalizes rational forms of activity organization that have been verified in rich social and historical practice.** During different epochs of civilization development, various **basic types of organizational forms of activity** have been popular. In modern scientific literature, they are often referred to as *organizational culture* [1, 9, 14, 25, 42].

For instance, V.A. Nikitin (see references in [29]) identifies the following historical types of organizational structure (see Table 1.1). Let us discuss them in a greater detail; this would be useful for further exposition.

Traditional organizational culture. In the early stages of mankind's development, a society consisted of communities, where differentiation was based on the kinship principle. Communities were maintained by *myths* and *rituals*. A myth can explain the origin of ancestry (e.g., from an animal or a god), the peculiarity of a given group, the rules of communal life (in particular, the primacy principle in a group and its

Table 1.1 The types of organizational culture (characteristics by V. Nikitin).

The types of organizational culture	The methods of normalization and translation of activity	The forms of social structure implementing the corresponding method
Traditional	Myths and rituals	Communities based on the kinship principle
Corporate-handicraft	Samples and recipe for their recreation	Corporations with a formal hierarchical structure (masters, apprentices, and journeymen)
Professional (scientific)	Theoretical knowledge in the form of text	Professional organizations based on the principle of ontological relations (relations of objective reality)
Project-technological	Projects, programs[2] and technologies	Technological society being structured by the communicative principle and professional relations

substantiation). A myth can define the structure of the world, i.e., separates another world ("the next world," the world of spirits, and so on). The latter resembles the real world, yet possesses supreme and perfect qualities against the real counterpart. The life in a community took place in both worlds simultaneously. The real mechanism, which ensures such correlation and organizes human activity, is provided by a ritual. The primary task lies in separating aliens from relatives, helping the latter and being injurious to the former, as well as in punishing for apostasy.

Corporate-handicraft culture. In the 6th century, a new social structure, with the rigid hierarchy of the Church, gradually substituted communities; this process was going on under the active influence of the Roman Empire. The Church had higher corporate organization, *viz.*, a unified control authority and a common ideology, a clear hierarchy of subordination, an internal system of education (personnel training), explicit norms of behavior and punishment for disobedience and a common language (Latin).

The Late Middle Ages were remarkable for the appearance of new centers of society organization – cities and universities. The new social hierarchy within cities was formed involving alternative (in fact, corporate-handicraft) principles. Corporations concentrated on a specific activity. Notably, some samples (e.g., of products) and recipes for their recreation were prepared and carefully protected by a corporation. The hierarchical structure of society was subject to a fixed differentiation of the members of handicraft corporations (masters, apprentices and journeymen). Transition between categories required time and had many conditions, controlled by a corporation.

During the Renaissance, university corporations gradually substituted the application of recipe knowledge for the application of theoretical knowledge. Accordingly, definite interest arose in the people being able to create *theoretical knowledge* and transmit it (instead of the corresponding recipe knowledge). Transmission of theoretical knowledge became the key aspect for universities and (later) for other forms of education. Thus, the professional type of organizational culture started its formation.

---

[2]In the current sense, programs represent large-scale goal-oriented projects.

The professional (scientific) type of organizational structure. Here the basic activity cementing different professional fields is represented by *science*. In a professionally organized society, science makes up the major institution; indeed, it serves for forming a unified structure of the world and general theories (afterwards, specific theories and corresponding problem domains of professional activity are separated with respect to the unified structure of the world). The "center" of professional culture lies in scientific knowledge, while generation of such knowledge represents the major type of production (affecting the capabilities of other types of material and immaterial production). The professional type of organizational structure was the leading one within several centuries.

However, in the second half of the 20th century, cardinal contradictions were observed in the development of the professional form of social structure. They were:

- contradictions in the unified structure of the world suggested by science, and internal contradictions in the structure of scientific knowledge generated by science, the beliefs about shifts of scientific *paradigms* (T. Kuhn [20], K. Popper [37], I. Lakatos [21] and others);
- onrush development of scientific knowledge, "technologization" of the means to generate scientific knowledge resulted in diversification of the world structure (leading to fragmentation of professional fields into numerous specialities).

Therefore, there was an immediate necessity to develop another type of organizational structure, *viz.*, the project-technological one.

The project-technological type of organizational culture. As far back as the previous century, many theories were accompanied by new structures such as *projects* and *programs* [1, 25]. Moreover, by the end of the 1990s the activity regarding creation and implementation of projects and programs became very popular. These structures are supported by analytical work rather than by theoretical knowledge. Due to its theoretical strength, professional culture generated certain ways of mass production of new sign forms (models, algorithms, databases, etc.) – the "fabric" for new technologies. The above-mentioned technologies serve not only for material production, but also for sign production. Generally speaking, *technologies* (in addition to projects and programs) became the leading form of activity organization.

We have provided merely one of numerous classifications used for historical types of *organizational culture*[3] [9, 14, 42]. Alternative approaches can be found in scientific literature. The most important aspect consists in the following. The professional type of organizational culture based on written texts (handbooks, manuals, instructions, procedural recommendations) had been gradually developing since the 17th century. Meanwhile, around the 1950s, it was replaced by a new type of organizational culture (naturally, the new one absorbed the previous types), *viz.*, by project-technological

---

[3]In many sources, the notion of organizational culture is used in a narrower sense (as the culture of organizations or corporate culture). Corporate culture is the mission of an enterprise (an organization, etc.), its organizational structure, the system of norms, traditional internal relations, symbols, and so on.

culture[4]; this process was induced by rapid development of social (including industrial) relations.

Let us emphasize another feature. As completed cycles of the productive (creative) activity, both performing scientific research and making a work of art fit the stated definition of a project. In science and art, the term "project" has been adopted recently (starting from the 1950s, e.g., an atomic project, a movie project, a play performance project). However, the project type of organizational culture was first mastered by painting – in the Renaissance, art was separated from handicrafts due to the formation and development of the category of an image as the artistic model of reality. This process took its final shape by the beginning of the 19th century (in particular, we refer an interested reader to *Aesthetics* by G. Hegel).

At the confine of the 19th and 20th centuries, the project type of organizational culture "penetrated" into science. In many fields of scientific knowledge, the requirement appeared concerning formation of *scientific hypotheses* as cognition models. In fact, a scientific research was organized in the form of projects. One would observe the fully-fledged "operation" of the project-technological type of organizational culture merely in recent decades – it has been widely demanded by the practice.

The new type of organizational culture discussed above includes the following key notions: a project, a technology, and reflexion. Yet, the first and the last ones are somewhat contrary – a project (*verbatim*, "sent forward") and reflexion (*verbatim*, "addressing back").

We consider these notions in a greater detail. An old traditional interpretation of a project (e.g., in engineering, construction) consists in the totality of documents (calculations, drawings, and so on) to design a building or a product. Later on, it was substituted by the modern conception of a project as a completed cycle of the productive activity (performed by an individual, a collective, an organization, an enterprise, or by several organizations and enterprises).

A *project* is a purposeful creation or modification of a certain system, having a specific organization under constraints imposed on available time, resources and quality of the results.

The presence of a certain system in the above definition indicates the project's integrity, singleness and uniqueness, as well as its features of novelty.

There are numerous projects to-be-faced in real life. They vary in the aspects of problem domain, application, scale, duration, staff, complexity, and others.

For comfortable analysis of projects and project management systems, one may classify projects using different bases, as follows.

Project type (according to the scope of activity of a specific project): technical projects, organizational projects, economic projects, social projects, educational

---

[4]We underline that the types of organizational culture do not simply replace each other during their development. The matter is much more complicated, since different types of organizational culture coexist. For instance, many ceremonies and rituals have been permanently in a nationality since ancient times (e.g., Russians mostly profess Orthodoxy and still have heathen feasts such as Maslenitsa). Another example is that the activity of some modern scientific schools is organized according to the corporate-handicraft type of organizational culture. Furthermore, certain kinds of human activity can be based on different types of organizational culture.

projects, investment projects, innovation projects, research projects, training projects, mixed-type projects.

Project class. The following classes of purposeful changes are identified depending on the scale (in the ascending order) and on the level of interdependence:

- *works* (operations);
- *batches of works* (the complexes of technologically interrelated operations);
- *projects*;
- *multiprojects* (a multiproject is a project which consists of several technologically related projects with shared resources);
- *programs* (a program is the complex of operations (measures, projects) with technological, resource and organizational interrelations, ensuring a required goal [1]);
- *project portfolios* (in the general case, the set of technologically independent projects, being implemented by an organization under certain constraints and ensuring its strategic goals).

To describe the above-stated elements, one should account for goals, resources, the technology of activity, and control mechanisms. Each of these aspects defines the corresponding class of purposeful changes:

- in the case of a multiproject, of crucial importance are technological constraints (imposed on the interrelation of the embedded works and subprojects) and resource constraints;
- in the case of a program, of crucial (backbone) importance is ensuring a given goal under existing resource constraints;
- in the case of a project portfolio, of crucial importance is using unified control mechanisms that ensure the most efficient attainment of strategic goals of an organization under existing resource constraints (a project portfolio is always associated with an organization implementing it).

Project duration (according to the period of implementation of a project): short-term projects (below 3 years), middle-term projects (between 3 and 5 years), and long-term projects (above 5 years).

Project complexity (the level of complexity): simple projects, difficult projects, and extremely difficult projects.

Involving the fundamental concept of a project, we may consider scientific research as the form of projects, i.e., as completed cycles of scientific activity.

Each project passes a series of development stages (starting from idea initiation to its total completion). The whole set of development stages makes up a *life cycle* of a project. Traditionally, a life cycle is decomposed into *phases*, phases are decomposed into *stages*, and stages are decomposed into *steps* [1, 29].

To avoid confusion, we make a clear provision regarding the difference between the notions of a project and design. *Design* is the initial phase of any project.

Indeed, any *productive activity* and any project require specific goal-setting (i.e., design). Any scientific research is designed, as well.

Now, let us proceed to the next definition ("technology"). Its modern interpretation lies in the following. A *technology* is a system of conditions, forms, methods and means to solve a posed problem. Such understanding of a technology has been recently imported from the industrial sphere. This process was initiated when in developed countries know-how engineering companies (companies designing new types of products, new materials, new processing techniques, etc.) started forming independent structures. These companies sold licences for production of their developments to vendors; such licences were accompanied by a detailed description of manufacturing means and techniques (i.e., technologies).

Naturally, any project is realized by a set of technologies.

An essential role in organization of the productive activity is played by *reflexion* as permanent analysis of goals, tasks, and results of the process.

Similarly to the methodology of other types of human activity, research methodology can be constructed in the logic of project category based on the triad of **project phases**:

- DESIGN PHASE;
- TECHNOLOGICAL PHASE;
- REFLEXIVE PHASE.

Each phase includes particular stages and steps.[5]

Therefore, we have studied the basic philosophical, psychological and systems engineering notions being necessary for further exposition. Next section analyzes epistemological foundations of methodology.

1.2   EPISTEMOLOGICAL FOUNDATIONS

Methodology (as the theory of organization of an activity) naturally rests upon *scientific knowledge*. A researcher involved in scientific activity must have a clear and conscious conception of science, its organization, the laws of science development, and the structure of scientific knowledge. In addition, a researcher must conceive the criteria of scientific knowledge (for a new knowledge to-be-obtained as the result of investigations), as well as the forms of scientific knowledge to-be-used for expressing the results of investigations. That is, a researcher must understand distinctly the "footholds" of his/her scientific activity to make it meaningful and well-organized.

These issues are discussed in the present section.

The field of science studying science itself (in the general interpretation of this term) is called the *science of science*. Actually, it includes several disciplines such as epistemology, the logic of science, semiotics (the theory of signs), the sociology of science, the psychology of scientific creation, and others.

---

[5]For instance, design phase consists of four stages (conceptual stage, modeling stage, design stage and technological preparation stage). Next, modeling stage has the following steps: model construction, optimization, choice (see the details in [29]).

In the context of this book, of crucial importance is epistemology; in particular, *science methodology* (the methodology of scientific research) is often viewed as a branch of epistemology.

Epistemology is the theory of scientific cognition, a branch of philosophy. Generally speaking, epistemology studies the laws and capabilities of cognition, as well as analyzes the stages, forms, methods, and means of cognition process, the conditions and criteria of scientific *knowledge validity*. The general methodology of science as the theory of organization of scientific activity represents the branch of epistemology which focuses on the process of scientific activity (its organization).

Recall that methodology is the theory of organization of an activity, and scientific activity is organized according to specific complete cycles. Indeed, it seems impossible to develop science "on the whole"; an independent investigator or a group of scientists conduct a definite research (a scientific project), and switch to another one (a new project) as soon as the current research is finished. Therefore, the notions of the methodology of science, the methodology of scientific activity and research methodology are synonyms in a certain sense.

Moreover, we have to distinguish between the terms of *scientific cognition* and *scientific research*. Scientific cognition is considered as a sociohistorical process and represents the subject of *epistemology*. Scientific research makes up a subjective process – the activity regarding acquisition of new knowledge by an individual (a scientist, an investigator) or by a group of researchers; this is the subject of the *methodology of science* (the methodology of scientific activity, research methodology). Scientific cognition is part and parcel of cognitive activity of individuals; yet, the latter can cognize (study) a certain phenomenon as far as they possess the common (collective) objectified *system of knowledge*, being passed from one generation of scientists to another.

Thus, we have finished the brief terminological excursus. Now, let us consider epistemological foundations of methodology.

The general notions of science. The following opposite delusions are wide spread among many people who have little to do with science. On the one hand, many adhere to the opinion that science represents something mysterious, enigmatic, and accessible merely to a selected "handful." On the other hand, we should mention derogatory remarks about science and scientists; they are often considered as "bookworms, rummaging unnecessary things" (in contrast to practicians, who is "doing real work").

Both viewpoints turn out to be absolutely wrong. Similarly to any activity (e.g., teaching, production), science is a field of professional human activity. Perhaps, the only specific feature of science is that it serves to obtain scientific knowledge, whereas other fields of human activity utilize knowledge accumulated by science.

Science is defined as a field of human activity, whose function consists in generation and theoretical systematization of objective knowledge regarding the reality.

In a narrower sense, the term "science" indicates specific branches of scientific knowledge (e.g., physics, chemistry, psychology, pedagogics are sciences).

Science represents an extremely multi-aspect phenomenon. In any event, one should account for (at least) three basic aspects of science (making distinctions among them in a concrete case):

–   science as a social institution (the community of scientists, the totality of scientific establishments and structures of scientific service);

– science as a result (scientific knowledge);
– science as a process (scientific activity).

The first and second aspects will be analyzed in the present section. The third one – science as a process (scientific activity) – is connected with research methodology (see the next chapters of the book).

Science as a social institution. This is a large sector of a national economy. For instance, in the former USSR approximately 2 500 000 employees were engaged in the sphere of science and scientific service; thus, the country was ranked the first in the world by the number of scientific employees. Today, the state system of scientific establishments[6] includes hundreds of institutes and centers of *the Russian Academy of Sciences*, as well as sectoral scientific institutes. The primary structural units of scientific institutes and centers are research departments, laboratories, sectors and groups (here they are listed in descending order of the number of employees). Scientific establishments also embrace numerous technological and project institutions, engineering offices, scientific libraries, museums and national parks, zoological gardens and botanic gardens. In recent years one would observe the popularity of the so-called *science parks* as the unions of small self-supporting theoretical and practical firms; they perform scientific research using the base of large universities or industrial enterprises and implement the results by selling new technologies.

In any country, most of scientific potential is always accumulated by institutions of higher education. On the one part, the reason is that ensuring high-level training in an institution requires highly skilled research and educational personnel. On the other part, such an approach enables involving young students in research. Higher school establishments (universities, academies and institutes) have several hundred, or even thousand, members of professional and teaching staff (depending on the number of students). The basic pedagogical and scientific unit of an institution of higher education lies in a department (a chair).

It seems impossible to conduct research without an appropriate infrastructure. The latter includes the so-called agencies and organizations of scientific service (such as scientific publishing houses, scientific periodicals, libraries, scientific device design, etc.) – the subbranch of science as a social institution.

Furthermore, science as a social institution operates merely under the presence of highly skilled scientific staff. The training of scientific staff is organized in graduate schools (at the level of candidate of science degree, PhD, PostDoc programs, etc.).

Next, the institution of doctoral candidacy serves to prepare scientific staff of the highest level (doctors of science).

In addition to scientific degrees, lecturers in higher schools are assigned academic ranks depending on their pedagogical levels. These ranks include associate professors (prerequisites: preferably candidates of science with sufficient teaching experience in higher school and scientific publications) and professors (prerequisites: preferably doctors of science with important scientific works such as numerous recognized papers in leading journals, textbooks, monographs, etc.).

---

[6]In addition to state academies, there exist many public academies of sciences.

Science as a result. In this sense, *science* is defined as the system of authentic knowledge about nature, man and society. It seems relevant to emphasize the following aspects in the above definition:

1   Science as the system of knowledge – here science must be treated as an interrelated set of *knowledge* regarding all current issues of mankind (nature, a man and society); such set must satisfy the requirements of completeness and consistency.
2   The matter concerns only *authentic knowledge* (in contrast to ordinary-practical knowledge and the beliefs of an individual). *Scientific knowledge* is a specific form of reflection of the reality in human minds. The other forms of reflection include *art, religion,* and *philosophy.* Science acts in the following links with them:

- science – art: science operates with notions, while art involves images;
- science – religion: science operates with knowledge, while religion is based on faith;
- science – philosophy: science operates with knowledge, while philosophy deals with general views of the reality (philosophy is based on scientific knowledge and is a field of science).

General laws of science development. There exist six general laws of science development (see references in [29]).

1   The conditionality of science development on the needs of sociohistorical practice. This is the major driving force or source of science development. Note that science development is subject to the needs of sociohistorical practice exactly (not just the needs of industrial or educational practice). A specific research can be unrelated to concrete needs of practice (simply being implied by the logic of science development or being defined by personal interests of an investigator).
2   The relative independence of science development. Whatever scientific problems are posed by practice, solving them is only possible when science reaches a corresponding level of development, a corresponding phase in the process of reality cognition. Moreover, sometimes a researcher must display enough courage when his/her scientific views are in conflict with established traditions, opinions of colleagues, directions of a certain ministry or active norms, documents, and so on.
3   The continuity of scientific theories, ideas and concepts, methods and means of scientific cognition. Each higher phase in science development proceeds from the preceding phase (the most valuable knowledge accumulated earlier is preserved).
4   The alternation of the periods of smooth (evolutional) and on-rush (revolutionary) development – breaking old theories, systems of concepts and beliefs. Evolutional science development is the process of gradual accumulation of new facts and experimental data within the framework of existing views (frameworks, dominating paradigms). Thus, one observes expansion, correction and refinement of the theories, concepts and principles adopted earlier. Revolutions in science are remarkable for radical changes in generally accepted laws, revision of fundamental principles and concepts via accumulating new data, exploration of new phenomena contradicting the previous views. However, scientists reject not the previous content of knowledge, but its wrong interpretation (e.g., incorrect universalization of laws and principles having a relative or limited applicability).

5   Interaction and interconnection of all fields of science; hence, a subject of a certain science can and should be analyzed by the techniques and methods of another science. As a result, necessary conditions are ensured for complete and deep disclosure of fundamentally different phenomena.

6   The freedom of criticism, unimpeded discussion of scientific issues, open and free expression of different opinions. The dialectically contradicting character of phenomena and processes in nature, society and human beings is revealed gradually and indirectly. And so, the opposing opinions and views merely reflect some contradictory aspects of studied processes. Such struggle makes it possible to overcome the initial (inevitable) unilateralism of different views concerning an object of research. As a result, a common view is generated, providing the most authentic reflection of the reality to date.

We mention the following **properties of science as a result:**

1   The cumulative character of scientific knowledge development. A new knowledge is united and integrated with an old knowledge (not rejecting, but supplementing the latter). In recent decades the development of scientific knowledge obeys the exponential law (i.e., the amount of scientific knowledge is doubled every ten years). Moreover, a new scientific knowledge can be obtained only if a researcher has carefully studied the outcomes of his/her predecessors. This is of crucial importance; indeed, sometimes practicians start "experimenting" without a detailed analysis of scientific literature on the subject (thus, "re-inventing the wheel" or "discovering America").

2   Science differentiation and integration. Accumulation of scientific knowledge causes differentiation (separation) of sciences. New fields of scientific knowledge appear (chemical biophysics and physical biochemistry, pedagogical psychology and psychological pedagogics, to name a few). At the same time, integration processes can be identified, as well; general theories are suggested, uniting and explaining hundreds or thousands of facts (one would think disconnected facts). For instance, D. Mendeleev's periodic law provided a comprehensive theoretical base to explain thousands of chemical reactions. On the other hand, J. Maxwell's equations of electromagnetic field explained all known phenomena of electricity and magnetism of that time. Furthermore, the equations made it possible to predict the existence of radiowaves and other phenomena.

The structure of scientific knowledge. Scientific knowledge is structured according to specific fields of science – see Fig. 1.2 and references in [29]):

–   the *central field* of scientific knowledge: physics, chemistry, cosmology, cybernetics, biology, anthropological sciences, social sciences, and technical sciences;
–   *philosophy*, representing simultaneously a field of science and the system of views of the reality; thus, it occupies a particular position (see the discussion above);
–   *mathematics*, occupying a particular position as a separate field of scientific knowledge (its subject consists in designing formal models of phenomena and processes studied by other sciences);
–   practical sciences (also called activity-related sciences): medicine, pedagogics, engineering sciences, and *methodology*.

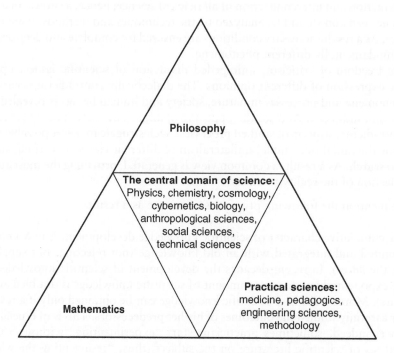

Figure 1.2 The structure of scientific knowledge.

We will not discuss various classifications for structures of scientific knowledge; such classifications are not the scope of this book. Let us consider the features of any field of scientific knowledge under conditions when different sciences appreciably vary in their epistemological level. Notably, there exist two camps; the first one lies in *epistemologically strong sciences* (including the epistemological ideal of science – mathematics, physics, other natural sciences whose theories are based on strict deduction).

The second camp is occupied by *epistemologically weak sciences, viz.,* the humanities and social sciences (due to the extreme complexity of their objects, weak predictability of phenomena and processes). See also the discussion of the "uncertainty principle" in Section 4.1). The following comparison seems quite appropriate here. After getting acquainted with the experiments of a great psychologist J. Piaget, an outstanding physician A. Einstein said that Piaget's theory of cognitive development in early childhood "is so simple that only a genius could have thought of it."

For each field of science, the following **attributes** are often identified:

1   Each field of science is related to a more or less separate set of cognition objects.
2   Given a set of cognition objects, there exist fixed relations, interactions and transformations making up the object of the corresponding field.
3   An object includes a relatively bounded range of problems being "clear" to experts. As cognition processes evolve, the range and content of these problems may vary (yet, preserving its succession). Meanwhile, one would definitely find "backbone"

problems being identical for all development stages of a given field and ensuring its self-identity.

4 There are validity criteria adopted within a given field of cognition.

5 Research methods used in a given field of cognition serve for solving rationally posed problems, agree with the chosen validity criteria and are directed towards the subject and object of knowledge in this field.

6 There exists an initial empirical basis of knowledge, i.e., certain information obtained as the result of a direct and indirect (perceptional) observation.

7 There is theoretical knowledge being specific for a given field of cognition (see the discussion below); one must not identify this knowledge with the notion of a theory, being used in the definition of an epistemological ideal of science (i.e., the matter by no means concerns theories in mathematics and physics). Generally speaking, theoretical knowledge does not necessarily act as a strict *deductive system*. Such knowledge cannot be expressed by formal mathematical calculus. Moreover, in contrast to rigorous theories (see below) that include merely logically interconnected laws, theoretical knowledge interpreted in a general sense, the above-mentioned knowledge contains concepts, hypotheses, principles, conditions and requirements, whose feature lies in the absence of an empirical origin. In particular, this applies to social sciences and the humanities.

8 There exists no separate, formal, *artificial language*, being specific only for a given field of knowledge. Nevertheless, one may speak about partial professional conceptualization, i.e., about partial changes in meanings and interpretations of terms, their adaptation to solution of problems in the system of professional research. For a long time, many fields of cognition have been utilizing a *natural language*, merely modifying its vocabulary. Their language differs from the common counterpart in its conceptual vocabulary, by not in its specific structure; the latter takes place for the fields related to the strong version of science.

The listed group of attributes can be called the weak or wide version of science. The term "weak" causes no emotional associations. It just fixes the existing situation, when some fields of scientific cognition do not fulfill the requirements of the strong version (the epistemological ideal of science which was established in certain historical conditions; it fixes a certain level of scientific development).

Now, consider the disciplines corresponding to the weak version of science in the historical perspective (taking into account their development trends). Consequently, one would clearly note that these disciplines have been gradually approaching the epistemological ideal.

In former times, the disciplines falling under the strong version did not fully satisfy the requirements of the epistemological ideal of science. Those disciplines occupied the stage being taken up today by some groups of the weak version of science.

The criteria of scientific knowledge. The issue regarding the *criteria of scientific knowledge* appears relevant to any science and any scientific research. Notably, what attributes should be used to separate scientific knowledge from the whole set of knowledge (including unscientific forms of knowledge)? Different authors suggest different approaches.

Let us provide the minimal set of attributes of scientific knowledge (see references in [29]). They are validity, intersubjectivity, and systemacy.

Knowledge validity. Knowledge validity is understood as its correspondence to a cognizable object – any knowledge must be object-oriented, since there is no knowledge "about something." However, validity is inherent to other forms of knowledge, e.g., to prescientific and ordinary-practical knowledge, as well as to judgements and conjectures, etc. Epistemology discriminates between the notions of validity and knowledge. The former implies the correspondence between knowledge and the reality, authenticity of its content irrespective of a cognizing subject (due to its objectivity, the content of knowledge exists independently of the subject). The notion of knowledge expresses the form of acknowledgement, presupposing the presence of certain grounds; depending on how such grounds are sufficient, one can speak about different forms of acknowledgement, *viz., judgement, trust, ordinary-practical knowledge,* or *scientific knowledge.*

However, scientific knowledge is remarkable for the following. The validity of certain content is always accompanied with *foundations* used to establish such validity (e.g., the results of an experiment, the proof of a theorem, a logical inference, etc.). Therefore, as an attribute characterizing the validity of scientific knowledge, one often indicates the requirement of its sufficient soundness (in contrast to insufficient soundness of other forms of knowledge). Hence, the *principle of sufficient reason* ("the law of sufficient reason" in *logic*) provides a basis for any science. Any valid idea must be substantiated by other ideas whose validity has been shown earlier. This principle was stated by G. Leibniz; in its classic form, the principle of sufficient reason is simply *"nothing is without a reason"* (*nihil est sine ratione*) or *"there is no effect without a cause."*

Intersubjectivity. This attribute expresses the property of general meaning, universalism, the mandatory property of science knowledge for all people (e.g., in contrast to a separate judgement being individual and not general). In this case, the following distinction takes place between the validity of scientific knowledge and the validity of other forms of knowledge. The validity of ordinary-practical knowledge, the validity of trust and the others remain "personal," since they are connected with the forms of knowledge requiring acknowledgement based on insufficient grounds. Concerning the validity of scientific knowledge, one easily observes their universalism, "impersonality" and being based on acknowledgement according to objectively sufficient foundations. The attribute of intersubjectivity is concretized by the requirement of *reproducibility* of scientific knowledge, i.e., by the identity of the results obtained by different investigators for the same object under the same conditions. On the contrary, a knowledge being invariant with respect to any cognizing subject may not claim to be scientific (since the requirement of reproducibility is not satisfied).

Systemacy. Systemacy characterizes different forms of knowledge. It is closely connected with the orderliness of both scientific and art (ordinary) knowledge. The systematic orderliness of scientific knowledge is conditioned by its organization engendering no doubts in the validity of its content. Indeed, scientific knowledge possesses a rigorous inductive-deductive structure, the property of rational knowledge derived as the result of reasoning based on available experimental data.

Thus, we have emphasized that the specifics of scientific knowledge lies in three attributes, i.e., validity, intersubjectivity, and systemacy. Each attribute itself does not form science; validity can exist in unscientific forms of knowledge, a "general delusion" may be intersubjective, as well; finally, the systemacy attribute (being realized independently of the ones of validity and intersubjectivity) determines merely "a sciolistic character," a semblance of validity, and so on. Only simultaneous realization of all three attributes in a certain result of cognition completely defines the scientific character of knowledge.

Accordingly, any research must comply with the stated (so to say, "classic") criteria of being scientific. At the same time, any criteria and requirements appear to be relative – other approaches to scientific knowledge are applicable.

The classifications of scientific knowledge. Generally, different classification bases are involved for scientific knowledge:

– according to the groups of problem domains, knowledge is classified as mathematical, physical, humanity-type and technical knowledge;
– according to the way of reflecting its essence, knowledge is classified as *phenomenological* (descriptive) and *essentialist* (explanatory) knowledge. Phenomenological knowledge represents qualitative theories with *par excellence* descriptive functions (many branches of biology, geography, psychology, pedagogics and so on). Contrariwise, essentialist knowledge makes up explanatory theories with application of quantitative analysis tools;
– according to the activity of certain subjects, knowledge is classified as *descriptive* and *prescriptive*, normative knowledge; the latter contains regulations, direct instructions for an activity. We underline that the material regarding the science of science (in particular, epistemology) presented in this subsection has a descriptive character. Nevertheless, first, this material is necessary as a guideline for any investigator. Second, it provides a certain base for further exposition of prescriptive, normative material related to the methodology of scientific activity;
– according to functional purposes, scientific knowledge is classified as *fundamental*, *applied* and *development* knowledge;
– and so on (there exist numerous classification bases).

In the context of this book, of crucial importance is the classification of scientific knowledge according to the forms of thinking – distinguishing between empirical and theoretical knowledge.

Empirical knowledge is the established scientific facts, as well as the empirical laws formulated on their basis. Hence, an empirical research directly aims a certain object and involves empirical, experimental data.

Empirical knowledge represents an absolutely necessary stage of cognition, as far as all knowledge arises from experience. Nevertheless, such knowledge is insufficient for cognizing deep internal laws of origin and development of an object.

Theoretical knowledge is the general laws stated for a given problem domain, enabling to explain the facts and empirical laws established earlier, as well as to predict and foreknow future events and facts.

Theoretical knowledge transforms the results obtained at the stage of empirical cognition into deeper generalizations, thus revealing the essence of the phenomena at levels 1, 2, ..., the laws of origin, development and changes in an object considered.

To comprehend these distinctions, let us give illustrative examples. The well-known Ohm law is empirical. The matter concerns the gas laws of Boyle-Mariott, Charles and Gay-Lussac. At the same time, the Clapeyron-Mendeleev equation (the model of an ideal gas), which generalizes the above-mentioned gas laws based on the molecular-kinetic theory, makes up a theoretical knowledge.

The both types of knowledge – the empirical and theoretical ones – are inherently interconnected, affecting the development of each other within the integral structure of scientific cognition. By identifying new scientific facts, empirical research stimulates the development of theoretical research and poses new problems for the latter. On the other hand, theoretical research develops and concretizes new prospects of explanation and prediction of facts, thus directing and accompanying empirical research.

Historically, the empirical stage of science development (e.g., in the case of natural science, that was the period from the 17th century till the beginning of the 19th century) was remarkable for the following. The primary means of forming scientific knowledge consisted in empirical research and their subsequent logical generalization in the form of empirical laws, principles and classifications. Further development of the conceptual framework of science yielded the appearance of logical forms such as typology, primitive explanatory schemes, models, whose content lay outside the scope of the initial generalization and comparison of empirical data. The formation of integral theoretical systems signified the transition of science to the theoretical stage. The latter is described by the appearance of specific theoretical models of the reality, preconditioning the progress of theoretical knowledge rather independently of the empirical level of research. The development of the theoretical content of science and building of multilayer theoretical systems cause certain isolation of the theoretical framework from its empirical basis.

The dialectics of relations between empirical and theoretical knowledge is such that (early or late) the latter is constructed based on the former. For instance, in the author's statement, Kepler's laws of planetary motion were empirical generalizations. Following the advances in the classical mechanics, these laws were derived as corollaries of the Newtonian law of gravitation (obviously, this law possesses a fundamental character).

The forms of organization of scientific knowledge. This subsection serves as a certain vocabulary with references; the authors tend their apologies to the readers. However, the matter is that scientific literature provides almost no systematic treatment for the forms of organization of scientific knowledge. Therefore, we have considered it necessary to focus on a comprehensive analysis of this issue. Indeed, any research inevitably involves such forms, and many investigators use them in the "hit-and-miss" manner.

The result of science development reveals in scientific knowledge; hence, the latter must be expressed in definite forms. Let us list several forms of organization of scientific knowledge.

- a *fact* (also known as an *occurrence*, a *result*). A scientific fact includes only occurrences, phenomena, their properties, interconnections and relations being fixed or detected in a specific way. Facts make up the foundation of science. It seems impossible to construct an efficient scientific theory without a specific set of facts. A distinguished Russian physiologist, I. Pavlov, used to say that "facts give air for a scientist."

Facts as a scientific category differ from phenomena. A *phenomenon* is the objective reality, a separate occurrence, whereas a fact represents a collection of several phenomena and interconnections, their generalization. To a large extent, a fact is the result of generalizing all similar phenomena, combining them within a definite class of phenomena.

We should underline that scientific facts (even entering the structure of scientific theories) are independent of the theories; naturally, scientific facts are determined by the material reality. Therefore, scientific facts turn out invariant – certain theories can be disproven by practice, and the facts used to construct them go over to other theories. Nevertheless, the facts *per se* do not form sciences as knowledge systems. They perform their function only being embedded in the "fabric" of scientific knowledge, being within the frameworks of scientific theories. This idea was figuratively expressed by A. Poincaré: "The scientist must organize knowledge; science is composed of facts as a house is composed of bricks; but an accumulation of facts is no more a science than a pile of bricks is a house" [35].

- a *thesis* is a scientific assertion, a formulated idea. Particular cases of a thesis are an axiom and a theorem. An *axiom* is an initial thesis of a scientific theory taken to be valid without a logical proof and used to prove other theses of the theory. The issue regarding validity of an axiom is solved either within the framework of another theory or by means of interpretation, i.e., a meaningful explanation of this theory. A *theorem* is a thesis whose validity is established through a logical proof. Auxiliary theorems that serve to prove a basic one are called *lemmas* or *statements*;
- a *concept* is an idea reflecting (in the generalized and abstracted form) objects, phenomena and their interconnections by fixing general and specific attributes – the properties of objects and phenomena.

Scientists often speak about *evolving concepts*; notably, the content of a concept may acquire new attributes and properties as scientific data gets accumulated and scientific theories get developed.

Concepts occupy a particular position among other forms of organization of scientific knowledge. Actually, facts, theses, principles, laws, theories, etc. are expressed by words – concepts and their interconnections. Furthermore, the supreme form of human thinking lies in conceptual, verbal and logical thinking. According to G. Hegel, understanding means expressing in the form of concepts.

The process of concepts formation and development is studied by *logic* (*viz.*, by formal logic and by dialectical logic). *Formal logic* concentrates on the general structure of concepts, their types, the structure of defining the concepts, their structure within more complex contexts, and the structure of relations among concepts.

Dialectical logic analyzes the process of concepts formation and development in connection with transition of scientific knowledge from a less deep essence to a deeper one; dialectical logic considers concepts as the stages of cognition, as the result of scientific cognition activity.

The following structures related to concepts are treated in the logic of science: the content of a concept, the scope of a concept, the converse law between the content and scope of a concept, the division rules for the scope of a concept, specific and generic concepts, single and general concepts, concrete and abstract concepts, and so on. And finally, logic determines the seven fundamental *rules of concept definition*. Probably, their ignorance by investigators yields the definitions of concepts resembling the classic example of an incorrect concept: "A dog is an animal with a head, a tail and four legs." Really, such definition covers almost all terrestrial animals.

- a *category* is an extremely wide concept reflecting the most general and essential properties, attributes, interconnections and relations of objects and phenomena of the surrounding world (e.g., matter, motion, space, time, etc.). Each science possesses its own system of categories.
- a *principle* is a concept playing a dual role. On the one hand, a principle acts as a central concept representing the generalization and extension of a thesis to all phenomena and processes in a domain used to abstract this principle. On the other hand, it acts in the sense of an action principle – an activity norm, an activity instruction;
- a *law* is an essential, objective, general, stable and repetitive relation between phenomena and processes.[7] For instance, we refer to Ohm's law, the Joule-Lenz law, etc.

The surrounding world represents the set of material objects and phenomena having diverse complex interconnections and relationships. Hence, the most important relations (interconnections) between objects are posed as laws. Indeed, a law is exactly the essential relation being inherent not to a separate object, but to the whole set of objects making up a certain class, type, a set of uniform objects. The essential relation between objects, phenomena (or between their sides) defining the character of their existence and development expresses the major attribute of a law.

Generality is also an important feature of a law. Generality means that any natural or social law is intrinsic to all (without exception) objects and phenomena of a definite type or level, i.e., to the whole set of objects and processes described by this law. All material objects (from microscopic particles to giant stars) obey the law of universal gravitation. Similarly, all electrically charged bodies satisfy the law of electrostatic attraction derived by C. Coulomb.

Since a law is the essential (necessary) relation between objects (phenomena), it has a stable and repetitive character. Nevertheless, the stability of a law is not absolute – varying conditions may change or even destroy it. The essential interconnections describing the objective natural and social laws take place everywhere and all the time (but under the existing similar objects and proper conditions). Of course, the inverse assertion – repetitive interconnections and relationships are laws – appears illegitimate. Repetition may be random or not reflect the essential sides of a natural phenomenon. The repetition of a law is a necessary feature of a law (but not a sufficient one!). Nevertheless, exactly the repetition of a law under rather identical conditions is of crucial importance for science; the absence of such repetition would totally rule out the possibility of cognizing the surrounding reality.

---

[7]Roughly speaking, this definition applies to the strong version of sciences (admitting the reproducibility of results, repetition of phenomena and processes, and so on). Yet, in the weak-version sciences (the humanities and social sciences) a law rather has the character of a normative model.

- a *theory*. Generally speaking, the term "theory" is used in two senses. First, in the very common sense it represents the form of activity of a socially developed man, being intended for acquiring knowledge about natural or social reality and forming the collective activity of society (together with practice). Thus, the concept of a theory is equivalent to that of public conscience in the supreme and most developed forms of its logical organization. As the supreme product of organized thinking, it mediates any human attitude towards the reality and is the condition of truly conscious transformation of the latter.

Second (and this is the narrower sense we are interested in), a *theory* is the form of authentic scientific knowledge about a certain set of objects, representing the system of interconnected assertions and proofs and containing the methods of explanation and prediction of phenomena and processes in a given *problem domain*, i.e., of all phenomena and processes described by this theory.

In the last (narrow) meaning, the term "theory" is considered in two interpretations. First, in the context of the weak version of science (see the earlier discussion), a theory is a complex of views, beliefs, and ideas directed to explain phenomena, processes and their interconnections. Accordingly, one can substitute the word "theory" by that of *"conception"*. For instance, we mention the theory (conception) of problem teaching in pedagogics, the theory (conception) of personality in psychology, the conception of cultural dialogue suggested by M. Bakhtin, and so on. Second, in the context of the strong version of science, a theory is the supreme form of organization of scientific knowledge, providing a comprehensive image of essential interconnections in a given problem domain (i.e., the subject of this theory); e.g., the theory of relativity, quantum theory, etc. In the stated rigor meaning, the term "theory" is almost unused for social sciences and the humanities. This is due to the extreme mobility, variability, low predictability (or even unpredictability) of the phenomena and processes studied by these sciences; another reason consists in the infeasibility of introducing measurable quantitative characteristics for them.

The following primary components can be identified in the structure of a theory in the general (abstract-logical) form:

1   the initial empirical base of the theory, which includes the set of facts and conducted experiments (being fixed in the corresponding field of science); they have been given a certain description, but still wait for their theoretical interpretation;
2   the initial theoretical base of the theory, which includes the set of assumptions, postulates, axioms, general laws, and principles of the theory;
3   the logic of the theory, which includes the set of admissible rules of logical inference and proof within this theory;
4   the set of theoretically derived corollaries, theorems, assertions, principles, conditions, etc. with their proofs; this is the largest part of a theory, implementing the basic functions of theoretical knowledge (the "body" of the theory, its content).

The general logical structure of a theory finds different expressions in different types of theories. The first type (actually, a widest class of modern scientific theories) is represented by *descriptive theories*. Sometimes they are also referred to as empirical theories. The examples are the Darwinism (the theory of evolution by C. Darwin) in biology, I. Pavlov's physiological theory, as well as modern psychological and pedagogical theories. Such a theory directly describes a certain group of objects; its empirical basis often appears extensive, and the theory solves the problem of ordering the facts.

General laws formulated using this type of theories represent the extension of an empirical material. Such theories are stated in terms of standard natural languages involving technicalities that correspond to the studied problem domain. Actually, the rules of logic are not explicitly defined, and the correctness of proofs is usually not verified (except the experimental tests). Descriptive theories chiefly have a qualitative nature, thus being limited (in the sense of a quantitative description of a certain phenomenon).

The second type of theories lies in *mathematized scientific theories*, using the framework and models of mathematics (e.g., physical theories). Mathematical *modeling* (see below) serves to construct a special ideal object (a model) replacing a certain real object. The value of mathematized theories increases, since the corresponding mathematical models may admit not just a single interpretation but several interpretations (including the objects of a different nature); the only requirement imposed on such models is obeying the constructed theory. For instance, the same differential equation can describe the motion of a mechanical system or the current-voltage dynamics in a circuit (the so-called electromechanical analogy). Meanwhile, wide application of mathematical tools in a mathematized theory raises an intricate problem of interpretation (i.e., meaningful explanation) of formal results.

The justification problem for mathematics and other formal sciences promoted the development of the third type of theories – *deductive theoretical systems*. Apparently, the first system was Euclid's *Elements*, viz., the classic geometry based on the axiomatic method. The initial theoretical base of such a theory is stated from the very outset; subsequently, the theory is supplemented by those assertions being logically derived from the base. All logical tools involved in a deductive theoretical system are rigorously fixed, and all proofs are formulated according to these tools.

As a rule, deductive theories are constructed in terms of special *formal languages*, sign systems. Being very general, such theories pose the problem of results interpretation, which is the condition of transforming a formal language into scientific knowledge (in its proper sense).

Let us emphasize the following aspects as being relevant for further exposition. First, any scientific theory consists of interrelated structural elements (laws, principles, models, conditions, classifications, etc.). Second, any theory (irrespective of the type) includes in its initial basis a *backbone element* (or a group of elements). For instance, Euclid's geometry proceeds from five initial axioms (postulates). The backbone elements in classical mechanics and quantum mechanics are Newton's second and third laws and Schrödinger's equation, respectively. We will use the concept of a backbone element of a theory in the sequel.

- a *metatheory* is a theory which analyzes structures, methods, properties and ways of constructing scientific theories in a certain field of scientific knowledge.
- an *idea* (in the philosophical sense – as a sociohistorical idea; in contrast to its common meaning, e.g., "an idea has occurred to somebody's mind"). This is the supreme form of cognizing the world, not just reflecting the object considered, but being directed to its transformation. Thus, ideas in science not only summarize

the experience of the preceding development of knowledge, but also provide the base for synthesizing knowledge into a certain integral system and searching for new ways of problem solving. The development of an idea has two "vectors," notably, the development within science and the development towards practical realization of the idea. The examples of scientific ideas include the quantum idea in physics of the 19–20th centuries, the modern idea of education humanization, etc. A distinctive feature between an idea and a theory (a concept) lies in the following; in contrast to the former, the latter can be created by a single author and not become widespread. An idea must win the recognition of a society, a professional community, or (at least) of their substantial proportion.

- a *doctrine* is almost a synonym of a concept, a theory. This term is used in two contexts, notably, in the practical sense, when the matter concerns views with a tinge of scholastic property and dogmatism (the corresponding derivatives include "doctrinaire" and "doctrinairism"), and in the sense of a complex or a system of views, directions of actions that have obtained the normative character by approval from an official body (a government, a ministry, etc.). For instance, consider a military doctrine, the doctrine of housing and communal services development and others.
- a *paradigm* also acts in two aspects, *viz.*, as an example from history (including the history of a certain science used to justify or compare something) and as a concept, a theory or a model of a problem statement accepted as standard solution of research problems.

Moreover, we should mention here two specific forms of scientific knowledge. First, a *problem* as "the knowledge about the lack of knowledge" (the knowledge regarding the issues modern science is not aware of; yet, this deficient knowledge is necessary for science and its development or for practice or even for both). In mathematics, mechanics, theoretical physics, a certain analog of a problem lies in a *task*. For a subject (an individual, a group, a social community, a society), this concept reflects the necessity to perform a specific activity. Note the following phrases as being common among researchers in these fields of scientific knowledge: "to formulate a task," "to solve a task," "correct formulation of a task means the half of problem solution," and others. The second specific form of scientific knowledge is a *hypothesis* as a "conjectural knowledge." In the case of confirmation of a hypothesis, the latter becomes a theory, a law, a principle, and so on. Otherwise, it dwindles (loses its significance).

Since the terms "theory," "problem," and "hypothesis" are of crucial importance for further exposition of the book, let us discuss them in a greater detail.

The general concept of semiotics. *Semiotics* is the science studying the laws of designing and functioning of systems of signs. Naturally, semiotics is a foundation of methodology, since human activity and human communication require the generation of numerous systems of signs; using them, people can exchange different information and organize their activity.

Consider a certain message that can be transmitted by one person to another (thus, the former shares with the latter his/her knowledge about a subject or his/her attitude to a subject). The content of this message is comprehended by the receiver under an appropriate way of translation (enabling the receiver to reveal it). This is possible if (a) the message is expressed in signs carrying the corresponding meaning and (b) both the sender and the receiver identically understand the relationship between the *meaning* and *signs*.

Human communication appears versatile, people need many systems of signs. The underlying reasons are the following:

- the features of transmitted information (as a result, different languages may be preferable). For instance, we note the distinctions between scientific language and natural language, between art language and scientific language, etc.
- the features of a communication situation, making a certain language more convenient. For instance, natural language and gestural language are applicable in a private conversation; natural and mathematical languages serve for delivering lectures (e.g., in physics); the language of graphical symbols and light signals is used in traffic regulation, and so on;
- historical development of a culture, being characterized by successive extension of communication capabilities of people. As appropriate examples, recall modern ample opportunities of mass communication systems based on polygraphy, radio and television, computers, telecommunication networks, and so on.

To be frank, the issues regarding semiotics applicability in methodology (as well as in science and in practice) have been insufficiently studied to date. Yet, numerous problems exist here. For instance, most researchers in the field of social sciences and the humanities do not address the methods of mathematical modeling (even when this is possible and reasonable) – they do not operate the language of *mathematics* at the level of a professional user. Let us provide another example; today many investigations are performed "at the junction" of different sciences (e.g., pedagogics and engineering science). A great deal of confusion takes place when a researcher "mixes up" both *professional languages*. Nevertheless, the subject of any research (e.g., a doctoral thesis) must be unique, i.e., lie in a single problem domain (a single science). Consequently, a single language must be primary, while the other one can be auxiliary.

The given examples show the presence of many semiotic problems in methodology. And they require a solution.

1.3 ETHICAL AND AESTHETICAL FOUNDATIONS

Aesthetical foundations of methodology. *Aesthetical activity* – aesthetical components of activity – is inherent to an individual in any activity. Generally speaking, the specifics and functions of such activity consist in the following. Aesthetical activity is the field of free self-expression of a subject in his/her attitude towards the world. According to K. Marx, human beings (in contrast to animals) can produce using any type of measure (ideal) and can apply to an object an appropriate measure; thus, human beings create following the laws of beauty.

Aesthetical activity has the object- and spirit-oriented character. The subject of aesthetical activity can represent any real object, being available to direct perception or imagination. For instance, take art works containing aesthetical information; the products of rational activity, whose utilitarian purpose is accompanied with their aesthetical value; natural phenomena being separated from natural series (the ordering is subject to human activity) and entering into the context of aesthetical culture. Furthermore, the subject of aesthetical activity may include aesthetically neutral phenomena, whose value is actualized or confirmed during activity. Finally, the sphere of particular interest of aesthetical activity has always been the world of a man (the socio-historical process, social life of people, their behavior and the inner (spiritual) world).

Of special importance are the aesthetical foundations in labor as the basic form of human activity. Well-organized free labor, which consists of different types of work alternating with recreation, becomes the basic form of manifestation and development of creative, spiritual and physical strength of a man. The aesthetical rudiments in labor cause transformation of labor into the first vital requirement. Being directed towards satisfaction of material and spiritual needs, labor becomes really human; it forms a need whose free satisfaction provides enjoyment to a man (similarly to the delight perceived by an artist creating a painting).

The aesthetical components play an essential role in scientific activity. For a real researcher, investigations bring the greatest aesthetical pleasure (perhaps, resembling the delight of an artist or an actor). However, there is a fundamental difference between the results of scientific and art activities.

In particular, art works are purely personal. Each art work is inalienable of the author. For instance, Beethoven's famous *Ninth Symphony* would have never existed if he had not composed it. The situation slightly changes in science. Scientific results are personalized, as well; each scientific book, article, etc. has its author. Many scientific laws, principles and theories are assigned the names of their founders. Meanwhile, the following seems clear. Just imagine science without I. Newton, C. Darwin, A. Einstein, or N. Lobachevsky; most probably, the scientific results associated with the above names would have been obtained by other researchers. These results would definitely have appeared, since they objectively represent necessary stages of scientific development. Indeed, we can recall numerous facts from science history when the same ideas in different fields of science were independently established by various investigators.

As a rule, the distinction between science and art is explained by that the former provides a conceptual, logically relevant and partiality-free knowledge, while the latter appears emotional, visual, sensory, concrete, and so on. However, the personal sympathy of researchers is often used in scientific discussions and their emotions are as strong as the emotions of artists. One can indicate different roles of emotions in the processes of art search and scientific search, as well as in perception of art works and the products of scientific labor. The difference lies in that the emotional component is not accounted for in science and scientific results (yet, it *de facto* exists). Emotions originate from the personality of an investigator. Meanwhile, scientific material (including its ultimate result) is presented "on behalf" of an abstract subject; hence, emotions are either eliminated or must not be considered as an internal (relevant) component of research. In art emotions are inherent both to an artist and to an empathetic reader, a listener, a viewer. The emotional component is a general characteristic of an art subject. Art represents a personal reflection of the reality, whereas science acts as a detached and objective reflection of the reality.

Therefore, aesthetics directly relates to the methodology of science as the theory of organization of scientific activity (i.e., provides one of its foundations). Finally, we have to consider the last foundation – ethics.

Ethical foundations of research methodology. Since any human activity takes place in a society, it is naturally based (must be based) on *morality* and must be organized according to moral *norms*.

As is well-known, *moral culture* of a society is characterized by the level of assimilating the moral requirements (moral norms, principles, ideals, etc.) by society members, as well as by the level of their practical realization in the forms of actions and everyday behavior (exhibiting in the attitude of an individual to other people, the whole society, in his/her aims, life plan, value orientation, and so on).

In the common sense, morality makes up the comprehensive whole including moral consciousness, moral relations and moral activity. Morality is social in nature – it possesses a concrete historical foundation conditioned by certain public relations.

Moral culture acts as the value adoption of the surrounding world by a man. Ethical values are a unique regulating mechanism of relations between a society and an individual; they run through the whole activity of individuals, the whole system of interaction among them. Ethical values provide a concrete expression for many categories of morality (good, a duty, honor, conscience).

Moral regulation aims to ensure the social, class and group coordination of human activity. Hence, moral values become the standards of a proper behavior. As a standard of the proper, they form the base of moral assessments for the activity of the mass, groups and individuals, facts and occurrences. In the case of collisions (the acts of a deviant behavior), moral assessments are used by the dominating public opinion to direct individuals and groups towards the standards of a proper behavior.

Note that the moral guides of a society and an individual vary. The morality of a society cannot be reduced to the sum of moral guides of individuals; similarly, individual morality appears nonidentical to public morality. The relations of contradictory unanimity exist between a proper behavior (the one fitting moral requirements of a society) and an actual behavior (practical morality – the acts of people reflecting their level of moral development). Such relations may lead to moral collisions.

The structural standards of moral culture as an integral system are listed below.

- the culture of ethical thinking (the ability of using ethical knowledge, applying moral norms to a specific life situation, etc.);
- the culture of feelings;
- the culture of behavior (the ability of choosing one's own behavior, acting pursuant to moral principles and norms adopted);
- etiquette (regulating the form and patter of behavior).

Thus, moral culture is an essential side of all activity of a person, a people, a class, a social group, a collective (reflecting the operation of a concrete historical system of moral values).

In the sense of its content, the moral culture of a society provides a larger integral coverage for the established system of moral values and orientations than personal moral culture (here the components of the system are revealed with unique individual specifics). To a certain degree, and in an individual perspective, a person accumulates the achievements of the moral culture of a society in his/her consciousness and behavior. This assists a person in acting in a moral way in typical situations, as well as activates creative elements of his/her moral consciousness for selecting moral decisions in untypical situations.

The above-discussed levels of moral culture are closely interconnected. In many respects, the level of moral culture development in a society is determined by the perfection of moral culture of individuals. On the other hand, the richer the moral culture of a society is, the ampler the opportunities for perfecting individual moral culture are.

Here we should consider two specific aspects of ethics,[8] the so-called corporate ethics and professional ethics.

Corporate ethics is the code of written and unwritten *norms* of relationships among employees in an enterprise, a firm, an organization or an institution, that have been established as traditions or fixed in normative documents (regulations, job descriptions). Naturally, each manager and employee must follow them.

Professional ethics. In addition to universal and public ethical norms, certain occupations have professional ethical norms (e.g., pedagogical ethics, medical ethics – recall the famous Asclepiades' (Hippocratic) Oath). Of course, activity is organized here according to these specific ethical norms.

A separate issue concerns the ethical norms in professional scientific activity – the norms of scientific ethics.

The norms of scientific ethics. An independent question to be discussed consists in *scientific ethics*. The norms of scientific ethics have no rigorous formulation as certain established codes, official requirements, etc. However, they do exist and can be considered in two aspects – as internal ethical norms (in a community of researchers) and as external ethical norms (as a social responsibility of researchers for their actions and consequences).

In particular, the ethical norms of a scientific community were described by R. Merton, as far back as in 1942, as the totality of four basic values [27]:

- universalism: the validity of scientific assertions must be assessed regardless of race, sex, age, authority, academic titles or degrees of their authors. Therefore, science is *a priori* democratic. The results obtained by a famous investigator must be subjected to the same criticism and verification as the results derived by a novice researcher;
- communism: scientific knowledge must be the common heritage of mankind;
- disinterestedness: a researcher has to seek for truth without mercenary motives. Rewards and recognition should be considered as a possible consequence of scientific achievements, but not as an end in itself;[9]
- organized skepticism: each investigator bears responsibility for assessing the quality of results obtained by his/her colleagues; he/she is still responsible for using in the research the data obtained by other investigators (except he/she has verified the accuracy of these results). Notably, science requires respect for the preceding

---

[8]In principle, it is possible to study other ethical components, e.g., religio-ethnic or territorial ones.

[9]Meanwhile, there exists competition in science. Researchers strive for obtaining a new result faster than the others (on the one part) and separate researchers and scientific groups rival for grants and state orders (on the other part).

researchers. On the other hand, one should be skeptical about their outcomes. Recall the famous aphorism by Aristotle, "Plato is dear to me, but dearer still is truth."

In contrast to internal (professional ethics), internal scientific ethics is realized in relations between science and a society as social responsibility of researchers. This problem was almost not faced by investigators until the 1950s (the appearance of nuclear-missile weapons, genetic engineering, large-scale environmental disasters and other phenomena of scientific-technical progress). Today, we observe a growing responsibility of investigators for the consequences of their actions.

Therefore, in this chapter we have studied the foundations of research methodology. Now, let us address the methodology of science itself. It will be discussed in the following logic: characteristics of scientific research activity (Chapter 2), means and methods of scientific research (Chapter 3), organization of scientific activity (Chapter 4), and organization of collective scientific research (Chapter 5).

Characteristics of scientific research activity

2.1 FEATURES OF RESEARCH ACTIVITY

We start the analysis with the features of research activity.

When discussing the features of research activity, one should distinguish between *individual scientific activity* (as the research process carried out by a single investigator) and *collective scientific activity* (as the research process involving the whole community of investigators working in a given field of science, or as the research process performed by a group in an research institute, by an independent group or scientific school, etc.).

The features of individual scientific activity are:

1 An investigator must explicitly delimit the scope of his/her activity and define the objectives of his/her research. Similarly to any field of professional activity, science includes natural division of labor. Studying science "in the whole" seems impossible; thus, an investigator must single out a certain direction, pose a definite goal and gradually move towards its achievement. We will focus on the issues of research design below. For the time being, note that any scientific research possesses the following property. In his/her "scientific journey," any researcher encounters interesting phenomena and facts currently being of crucial importance *per se* (and so, attracting considerable attention). Meanwhile, a researcher risks being distracted from the backbone course of his/her work; indeed, a detailed analysis of secondary phenomena and facts may reveal new phenomena and facts (generally, this process is infinite). As a result, the research becomes "fuzzy," and no certain outcomes would be yielded. Let us forewarn: this is a typical mistake for most novice scientists! An important quality of an investigator lies in the ability of concentrating on the primary problem (the rest, i.e., "secondary" ones should be used at the level of their description in modern scientific literature).

2 The foundation of any research consists in the results obtained earlier. Prior to exploring a certain field of science or studying a certain problem, one should get acquainted with the state-of-the-art (notably, examine the work of one's predecessors).

3 An investigator must master the corresponding scientific terminology and construct a conceptual framework. The matter by no means concerns adopting a complicated language (many novice scientists wrongly believe that scientific content directly depends on the level of descriptive complexity). Contrariwise, a merit

of a real scientist is the ability of expressing intricate things in a simple language. Moreover, a distinguished boundary between common language and scientific language must exist. Notably, common language has no specific requirements to the accuracy of terminology. However, speaking about the same concepts using *scientific language* immediately raises the question, "What is the meaning of the concept?" Thus, in each concrete situation a researcher has to elucidate the sense of a concept.

In addition, different *scientific schools* may coexist. Each scientific school constructs its own conceptual framework, has its own "language." Imagine the situation when a young researcher takes a term in the interpretation of a certain scientific school, another term in the interpretation of the other school, and so on. This would lead to the total discord of terms, and the researcher would not create a new system of scientific knowledge (whatever he/she claims, such research lies within the scope of ordinary-practical knowledge).

4   The results of any research must (a) be in printed form or electronic form and (b) be published as a scientific report, paper, book, etc. The stated requirements are subject to the following reasons. First, only printed or electronic form enables presenting new ideas and results using a rigorous scientific language. Oral form makes this almost impossible. Furthermore, writing a scientific work (even a small paper) appears really difficult for a novice researcher – the statements that are easily presented orally may seem "unwritable." Thus, one observes the same difference as between common language and scientific language. Many logical imperfections are usually missed in oral speeches. On the contrary, any text requires a rigorous logical structure, which is much more arduous. Second, the goal of any research is obtaining a new scientific knowledge and bringing it to the notice of people. Suppose the knowledge is kept only in the mind of an investigator; thus, it becomes unavailable and useless.

Moreover, the numbers of scientific publications and citations are the productivity indicators for any scientist (though, formal indicators). Each scientist manages the list of his/her publications, and nowadays Internet provides ample opportunities for citations analysis.

The features of collective scientific activity are:

1   *The pluralism of a scientific opinion.* Any research is a creative process; thus, this process must be not "overregulated." No doubt, the research of a scientific group may and should be strictly planned. But a competent investigator has the right for a personal view to be respected. Any attempts of diktat, pressing a common opinion on everybody have never led to a positive result.

The existence of different scientific schools in the same field of science is subject to the objective necessity of non-coinciding views, opinions, and approaches. Later on, life and practice confirm or reject distinct theories or even reconcile them (e.g., recall vehement opponents R. Hooke and I. Newton in physics, or I. Pavlov and A. Ukhtomsky in physiology).

2   *Communications in science.* Any scientific research can be performed within a definite community of investigators. The underlying reason is that any (even the

most skillful) scientist has to discuss with his/her colleagues the ideas, established facts or theoretical constructs (for avoiding mistakes and delusions). Many young researchers say, "I would work independently, participating in discussions of significant results only." Yet, such approaches often fail. Indeed, "scientific adventures of a castaway" have never generated anything sensible. A researcher gets "digged" into the search; being disappointed, he/she leaves the field of activity. Thus, *scientific communications* are of crucial importance.

For any investigator, a prerequisite of scientific communication lies in direct or indirect contacts with his/her colleagues working in the same field. This can be achieved by attending dedicated conferences, seminars, symposia (direct or virtual communication) and reading of scientific literature – papers in printed and electronic journals, books, etc. (indirect communication). In both situations, an investigator makes reports or publishes papers (on the one part) and listens to the reports and reads the papers published by the colleagues (on the other part).

3   *The application of research results*[1] is the most important feature of scientific activity; the ultimate goal of science as a sector of a national economy consists in application of the obtained results in practice. Nevertheless, the widely held feeling (among people having little to do with science) that the results of any research must be implemented appears incorrect.

In the general case, the mechanisms of application may be different. The results of separate works are published as theses and papers; subsequently, they are generalized (and "shortened") in books and monographs as purely *scientific publications*. Next, they are included in textbooks for students in a more general and systematic form. Finally, the most fundamental results are combined in textbooks for schoolchildren.

In addition, not all research admits practical applications. Sometimes investigations serve for enriching the science proper and the arsenal of its facts, as well as for theoretical development of science. A certain "critical mass" of facts and concepts being accumulated, one observes qualitative leaps in implementation of scientific achievements in practice. A classic example is mycology (the branch of biology concerned with the study of fungi, mold in particular). For many decades, people have said, "Mold must be destroyed, not studied." In 1940 A. Fleming discovered the bactericidal properties of penicillum (black mold fungus). During World War II, penicillum-based antibiotics saved millions of human lives. One would hardly imagine modern medicine without antibiotics.

2.2 PRINCIPLES OF SCIENTIFIC COGNITION

Modern science follows three fundamental *principles of scientific cognition*, i.e., the principle of determinism, the principle of correspondence, and the principle of complementarity. Perhaps, the principle of determinism has the centuries-old history (though its interpretation undergone substantial changes and supplements at the turn of the 19th and 20th centuries). On the other hand, the principles of correspondence and

[1]This feature equally applies to individual and collective research activity.

complementarity were stated at the junction of the 19th and 20th centuries; they were connected with the development of new directions in physics (relativity theory, quantum mechanics, etc.). Among other factors, these principles conditioned the transformation of classical science of the 1800s–1900s into modern science.

The principle of determinism. Being a general scientific principle, it organizes knowledge construction in concrete sciences. In the first place, determinism acts in the form of causality as the totality of circumstances preceding a certain phenomenon and provokes it.

In other words, there is an interconnection of phenomena and processes when, under some necessary conditions, a phenomenon or a process (being the cause) generates another phenomenon or process (being the effect).

An essential shortcoming of classical determinism (also known as Laplace determinism) lay in being restricted by direct causality with a purely mechanistical interpretation; the objective nature of uncertainties was rejected, and the probabilistic relations were taken out of the determinism to be considered opposite to the material determination of phenomena.

The modern comprehension of the principle of determinism presupposes the presence of diverse entitative forms of interconnection among phenomena such that many of them represent relationships having no direct causal nature (i.e., one would not explicitly observe the moment when a certain phenomenon generates another). For instance, we mention space-time correlations, functional dependencies, etc. In contrast to the determinism of classical science, in modern science of crucial importance are uncertainty relations stated in terms of probabilistic laws, fuzzy set relations or interval uncertainty relations (e.g., see [30]).

However, in the final analysis all forms of real interconnections among phenomena are based on entitative causality; no phenomena exist beyond this causality (including random events whose aggregate serves for establishing statistical laws). Recently one can observe wide application of probability theory and mathematical statistics to investigations in social sciences.

The principle of correspondence. In its original form, the principle of correspondence was defined as an "empirical rule" expressing the natural connection (in the limit case) between nuclear theory based on quantum postulates and classical mechanics, as well as between special relativity theory and classical mechanics. Traditionally, there exist four mechanics, *viz.*, classical mechanics by I. Newton (corresponding to large masses, i.e., the ones appreciably exceeding elementary particle masses, and to small velocities, i.e., the ones being considerably lower than the velocity of light), relativistic mechanics – relativity theory by A. Einstein ("large" masses and "high" velocities), quantum mechanics ("small" masses and "low" velocities) and relativistic quantum mechanics ("small" masses, "high" velocities). These theories coincide in the corresponding limiting cases. In the course of further development of scientific knowledge, the validity of the principle of correspondence was demonstrated for almost all essential discoveries in physics (later on, in other sciences). Subsequently, it became possible to provide a generalized statement of this principle. Notably, the theories whose validity has been established experimentally for a certain scope (a set of phenomena) are not rejected by the appearance of new (more general) theories. Instead, they preserve their meaning for this scope as a limiting form (a special case) of the new theories.

The conclusions of the new theories in the scope of the old ("classical") theory turn into the conclusions of the latter.

Note that the principle of correspondence holds rigorously within the evolutionary development of science. Nevertheless, "scientific revolutions" are also possible, when a new theory rejects the preceding one, thus entirely substituting it.

In particular, the principle of correspondence implies the continuity of scientific theories. Let us focus the attention of investigators on the necessity of following the principle of correspondence. Indeed, in the humanities and social sciences today we face many research works (especially, performed by experts in "exact" sciences) endeavoring to create new theories, concepts, etc., being almost but not related to the preceding theories. New theoretical constructions are fruitful for science development; yet, if they are not connected with the existing ones, science would lose its integrity, while scientists would not understand each other.

The principle of complementarity. This principle appeared as the result of new discoveries in physics at the junction of the 19th and 20th centuries; during this period, it was found that a researcher studying an object introduces certain modifications in it (e.g., by a device used in the experiments). The principle of complementarity was first stated by N. Bohr: "Opposites are complementary." Notably, integrity reproduction for a phenomenon requires the application of mutually exclusive "complementary" classes of concepts during cognition. In physics this means that acquiring the experimental data about certain physical quantities is invariably connected with modifying the data about other quantities being complementary to the former (the narrow interpretation of the principle of complementarity). Complementarity serves for establishing the equivalence between the classes of concepts providing a complex description to contradictory situations in different fields of cognition (the general interpretation of the principle of complementarity).

The principle of complementarity considerably altered the system of science. Classical science operated as an integral system intended for (a) obtaining the set of knowledge in the final and completed form, (b) eliminating from the scientific context the impact of researcher activity and the means used by him/her, and (c) assessing the absolute validity of the knowledge included in the science fund. This situation was changed by the principle of complementarity.

Let us emphasize the following important aspects:

- embracing the subjective activity of a researcher by the scientific context modified the essence of knowledge subject. Instead of the "pure" reality, the subject of knowledge became a certain "section" of the reality defined in the light of accepted theoretical and empirical means and ways of reality cognition by a subject;
- the interaction between a studied object and a researcher (e.g., using devices) definitely leads to different levels of displaying the object's properties depending on the type of interaction with the cognizing subject (in different, often mutually exclusive conditions). This implies the legitimacy and equivalence of different scientific descriptions of the object (various theories concentrated on the same object or problem domain). Obviously, this is why Voland from M. Bulgakov's famous novel *The Master and Margarita* says, "All theories deserve each other."

Table 2.1 The comparison of two epochs of science development.

Attributes	The epochs of science development	
	Classical science	Non-classical science
1. Object	"Natural process" is separated irrespective of its analysis conditions.	Prohibition to consider objectness "on its own account" without the methods of its development. "There is no object without a cognizing subject."
2. Cognition method	Postulation of the mirror-type direct-obvious correspondence between knowledge and the reality (naïve realism).	Complementarity: conscious utilization of the groups of mutually exclusive concepts in investigations (observations, descriptions).
3. The relation towards empirical data	The empirical methodology of reaching the truth. Knowledge as direct generalization of experience.	Construction "irrespective" of the experience of conceptual schemes organizing and guiding the comprehension of experimental data.
4. Truth	Valid knowledge as reality (and not as an imperative).	Different viewpoints on a system are not reduced to a single viewpoint – the "Divine viewpoint" (common view on the reality) is impossible.
5. Scientific knowledge	Only a comprehensively substantiated knowledge in a certain thorough sense appears scientific. The presence of uncertainties is treated as insufficient substantiation (hypothetical character) of knowledge.	The absolute accuracy and rigor of knowledge are unachievable.

We underline that, according to the principle of complementarity, the same problem domain can be described by different theories. For instance, classical mechanics can be described not only by Newtonian mechanics, but also by W. Hamilton's mechanics, H. Hertz's mechanics, or K. Jacobi's mechanics. These approaches differ in the initial positions (the basic nondefinable notions – force, impulse, energy, etc.) [22]. Similarly, there exist numerous psychologies, *viz.*, gestalt psychology (the nondefinable notion of an image), behaviorism (the nondefinable notion of behavior) and so on.

Another example is that nowadays many socioeconomic systems are studied by means of mathematical modeling involving different branches of *mathematics* such as differential equations, probability theory, game theory, etc. Interpreting the results of modeling of the same phenomena and processes using different mathematical tools yields, in this case, close yet nonidentical results [30]. Generally speaking, the differences between classical science and modern (non-classical) science according to the stated three principles of scientific cognition can be combined in Table 2.1 (see also references in [29]).

For many years, the authors of this book have been concerned with the following question. Why are the stated three principles of scientific cognition so important? For instance, some investigators consider larger sets of principles. However, these three principles are universally recognized, and nobody casts doubts on them.

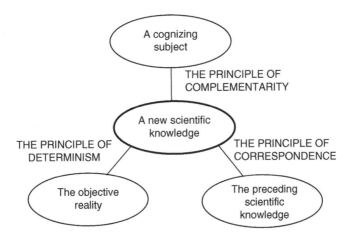

Figure 2.1 The logic of separating the principles of scientific cognition.

And the answer has been found! Indeed, it is easy. The goal of any research lies in obtaining a new scientific knowledge. This new knowledge relates to (see Fig. 2.1):

- the objective reality (the principle of determinism);
- the preceding system of scientific knowledge (the principle of correspondence);
- a cognizing subject, i.e., a researcher (the principle of complementarity); "there is no object without a cognizing subject."

Such an approach is efficient in explaining the principles of organization of scientific activity.

Therefore, in the present chapter we have analyzed the characteristics of scientific research activity. Now, let us pass to the means and methods of scientific research.

Chapter 3

Means and methods of scientific research

Means and methods are essential components of the logical structure of any activity. Consequently, they represent important subjects of research for methodology as the theory of organization of an activity.

Unfortunately, one would hardly find publications with a systematic treatment of the means and methods of an activity. They are discussed in many sources. Therefore, we provide a comprehensive description to this issue, striving to arrange the means and methods of scientific research in a uniform logic.

3.1 MEANS OF SCIENTIFIC RESEARCH

In the course of science development, the means of research (such as material, mathematical, logical and linguistic means) have been gradually designed and perfected. Obviously, recent times have demonstrated that we should also consider informational means as a particular class. All means of research are specially designed. In this sense, material, informational, mathematical, logical and linguistic means of research possess a common property – they are constructed, created, designed and substantiated for certain goals of research.

Material means of research include, in the first place, devices for scientific investigations. Historically, the appearance of material means of research was connected with the formation of empirical methods in investigations (e.g., observations, measurements, experiments).

These means are directed towards objects being studied, playing the principal role in empirical verification of hypotheses and other results of scientific research, in exploration of new objects and facts. Generally speaking, using material means of research in science (e.g., microscopes, telescopes, synchrophasotrons, Earth satellites) exerts a deep impact on forming the conceptual framework of sciences, on the ways of describing the objects being studied, on the techniques of reasoning and representation, as well as on the extensions, idealizations and arguments being involved.

Informational means of research. Large-scale implementation of computer technology, IT and telecommunication systems drastically changes research activity in many sciences, makes the latter the means of research, as well as enhances and simplifies scientific communications. In particular, recent decades have been remarkable for wide adoption of computer technology in experiment automation (in physics, biology, technical sciences, etc.). This enables appreciable simplification of research procedures and reduction of data processing time (by hundreds or even thousands times). Moreover, informational means allow for considerable simplification of statistical data processing almost in all sciences. The application of satellite navigation systems significantly improves the accuracy of measurements in geodesy, cartography, etc.

Mathematical means of research. Advances in mathematical means of research have a growing influence on the development of modern science. These means penetrate into the humanities and social sciences.

Mathematics represents the science of quantitative relations and spatial forms being abstracted from their specific content. Thus, mathematics has developed and utilized the concrete means of abstracting a form from its content, as well as has formulated the rules of form consideration as an independent object using values, sets, etc. This simplifies, facilitates and accelerates cognition process; moreover, this assists in deeper identification of the relationship among objects being abstracted from the form, as well as singles out the initial statements and guarantees the accuracy and rigor of judgements. Mathematical means enable considering not only directly abstracted quantitative relations and spatial forms, but also logically feasible ones (i.e., the ones derived by logical rules from known relations and forms).

Mathematical means of research cause essential changes in the theoretical framework of descriptive sciences. Mathematical means serve for systematizing the empirical data, identifying and formulating the quantitative dependencies and laws. Mathematical means are also used as specific forms of idealization and analogy (mathematical modeling).

Logical means of research. Any investigation poses several *logical problems* for a scientist, namely,

– what are the logical requirements for judgements to make objectively true conclusions? How can the character of such judgements be controlled?
– what are the logical requirements for the description of empirically observed characteristics?
– how can the logical analysis of initial systems of scientific knowledge be performed? How should certain systems of knowledge be agreed with other systems (e.g., in sociology and psychology)?
– how should a scientific theory be constructed for making scientific explanations, predictions, etc.?

Employing logical means in the process of constructing judgements and proofs aids a researcher in separating the controlled arguments from intuitively or not critically accepted ones, false arguments from true ones, confusions from contradictions.

Linguistic means of research. Important linguistic means of research include the rules of introducing definitions. Any research inevitably embraces refinements in proposed notions, symbols and signs, as well as application of new notions and signs. Definitions are always connected with a language as a tool of cognition and expression of knowledge.

The linguistic rules (in whatever languages – natural and artificial ones) used in constructing judgements and proofs, stating hypotheses and making conclusions, are the base of cognitive activity. Their knowledge exerts a considerable impact on the utilization efficiency of linguistic means in any research.

Means of scientific research are inseparably linked with methods of scientific research.

3.2   METHODS OF SCIENTIFIC RESEARCH

In any investigation an important (or even determinative) role is played by the *methods of scientific research.*

The methods of scientific research comprise *empirical* (*expressis verbis*, empirical means perceptible through sense organs) and *theoretical* ones (see Table 3.1).

The following aspect regarding the methods of scientific research should be outlined. Most publications on epistemology and methodology suggest double division (classification) of scientific methods, in particular, theoretical methods. For instance, the dialectical method, a theory (acting in the function of a method – see below), the identification and elimination of contradictions, hypotheses formation, etc. are often called the methods of scientific research (without proper explanations – at least, we have not found such explanations in literature). Meanwhile, such methods as analysis and synthesis, comparison, abstracting and concretizing (actually, the main mental operations) are frequently referred to as the methods of theoretical research.

A similar division takes place for empirical methods of scientific research. There exist partial methods (the analysis of publications, documents and results of activity; an *observation*; an *inquiry* (oral and written ones); *the method of expert evaluation*; *testing*) and complex or general methods involving a single or several partial methods (an investigation; monitoring; experience study and generalization; a trial; an experiment).

Such double division (both for theoretical and empirical methods) can be admitted by recalling the structure of activity.

We consider methodology as the theory of organization of an activity. If scientific research represents a cycle of activity, then the corresponding structural components are purposeful actions. As is generally known, an *action* is a component of activity whose distinguishing feature consists in the presence of a specific goal. At the same time, the structural components of an action represent operations, being correlated with the objective-subjective conditions of goal achievement. The same goal being correlated with an action can be achieved in different conditions; similarly, the same action can be implemented by different operations. However, the same *operation* may enter various actions (see A. Leont'ev [24]).

According to the aforesaid, let us single out (see Table 3.1):

– methods-operations;
– methods-actions.

The stated approach does not contradict the following definition of a *method*:

– first, a method as a way of attaining a certain goal or solving a concrete problem (a method-action);
– second, a method as a set of procedures or operations of practical or theoretical assimilation of the reality (a method-operation).

Table 3.1 The methods of scientific research.

Theoretical		Empirical	
Methods-operations	Methods-actions	Methods-operations	Methods-actions
• analysis • synthesis • comparison • abstracting • concretizing • generalization • formalization • induction • deduction • idealization • analogy • modeling • gedanken experiment • imagination	• dialectics (as a method) • scientific theories confirmed by practice • proof • the method of knowledge systems analysis • the deductive (axiomatic) method • the inductive-deductive method • the identification and elimination of contradictions • problems statement • hypotheses formation	• the analysis of publications, documents and results of activity • observation • measurement • (oral/written) inquiry • expert evaluation • testing	• the methods of object tracking: investigation, monitoring, experience study and generalization • the methods of object transformation: trial, experiment • the methods of object analysis in the course of time: retrospection, prediction

Therefore, in the sequel we will study research methods as

Theoretical methods:

–   methods – mental actions (identification and solution of contradictions, problem statement, hypothesis generation, etc.);
–   methods-operations (analysis, synthesis, comparison, abstracting, concretizing, and so on).

Empirical methods:

–   methods – mental actions (an investigation, monitoring, an experiment, etc.);
–   methods-operations (an observation, a measurement, an inquiry, testing, etc.).

Theoretical methods (methods-operations). Theoretical methods-operations have wide applicability, both in scientific research and in practical activity.

Theoretical methods-operations are analyzed with respect to three mental operations, *viz.*, analysis and synthesis, comparison, abstracting and concretizing, generalization, formalization, induction and deduction, idealization, analogy, modeling, a gedanken experiment.

Analysis is decomposing a studied object into several parts, identifying specific attributes and qualities of a phenomenon, process or relations among phenomena and processes. Analysis procedures appear an integral component of any scientific research, often forming its first phase (when an investigator passes from the whole description of a studied object to the identification of its structure, stuff, properties and attributes).

The same phenomenon or process can be analyzed in many aspects. The comprehensive analysis of a phenomenon leads to its detailed examination.

Synthesis is uniting different elements or sides of an object into the whole system. Synthesis means not just summing up the elements, but combining them in a certain sense. Indeed, simple unification of phenomena yields no system of interconnections among them (in this case, one merely has a chaotic accumulation of independent facts). Synthesis is opposite to analysis; yet, the both are inseparably linked. As a cognitive operation, synthesis acts in different functions of theoretical research. Any process of defining a notion is based on the unity of analysis and synthesis processes. Empirical data obtained in a certain research is synthesized at the stage of its theoretical generalization. In theoretical scientific knowledge, synthesis acts in the function of correlation of the theories belonging to the same problem domain, as well as in the function of combining of competitive theories (e.g., the synthesis of the corpuscular and wave theories in physics).

Synthesis plays a key role in any empirical research.

Analysis and synthesis are interconnected. Suppose that an investigator is more skillful in analysis than in synthesis. Probably, he/she takes the risk of being unable to find details in a phenomenon as the whole. Meanwhile, the relative prevalence of synthesis causes superficialism (paying no heed to relevant details may affect the comprehension of a phenomenon as the whole).

Comparison is a cognitive operation underlying the judgements regarding similarity or difference of objects. Comparison serves to identify quantitative and qualitative characteristics of objects, as well as to classify, order and assess them. Comparing means juxtaposing two objects. Of crucial importance are comparison bases or attributes determining possible relations between objects.

Comparison makes sense only within the set of homogeneous objects forming a class. The objects belonging to a certain class are compared using the principles being essential for consideration. Note that objects can be comparable by some attributes and noncomparable by the other. The higher is the accuracy of attributes' assessment, the better is the quality of phenomena comparison. An integral part of comparison lies in analysis, since any comparison requires separating the corresponding attributes of the phenomena being compared. Moreover, comparison is establishing certain relations among phenomena; naturally, for any comparison one should employ synthesis.

Abstracting is a basic mental operation, enabling mental separation of sides, properties or states of an object *per se*; thus, they make up a separate object of consideration. Abstracting underlies the processes of extension and definition of notions.

Abstracting consists in singling out object's properties that do not exist independently of the latter. Such procedure is feasible only mentally (as an abstraction). For instance, the geometric figure of a body does not exist on its own; the figure is inseparable from the body. However, due to abstracting one can separate and fix it, e.g., by a drawing. Thus, the specific properties of the geometric figure of a body are considered independently.

A major function of abstracting lies in separating the common properties of a certain set of objects and in fixing them (e.g., by introducing notions).

Concretizing is the opposite process to abstracting; it means identifying the whole, interconnected, versatile and complex entities. First, an investigator makes different abstractions; subsequently, by concretizing them he/she reproduces the integrity (the mental concrete) at a higher qualitative level of cognition. Therefore, in *dialectics* there exist two subprocesses in the cognitive process in terms of "abstracting-concretizing," *viz.*, ascending from the concrete to the abstract and ascending from the abstract to the new concrete (G. Hegel). The dialectics of theoretical thinking lies in the unity of abstracting (creating different abstractions) and concretizing (the motion towards the concrete and reproduction of the concrete).

Generalization is a basic mental operation which consists in identification and fixation of relatively stable invariant properties of objects and their relations. Generalization allows for reflecting the properties and relations of the objects irrespective of particular and random conditions of their observation. Using a specific viewpoint to compare objects belonging to a certain group, a man finds, identifies and denotes their identical (common) properties; the latter may form the content of the notion about this group or class of objects. Separating the common properties from the particular ones and denoting them enables the following. First, covering the complete variety of objects in a compact form; second, combining the objects in classes; and third, operating the notions without a direct reference to separate objects (via abstractions). The same real object may be included in narrow and wide classes; for this, one designs the scales of attributes similarity according to the principle of generic relations. The function of generalization lies in ordering the variety of objects and their classifying.

Formalization is reflecting the result of thinking in the form of exact notions or assertions. To a certain extent, this is a mental operation "having the second order." Formalization is opposite to intuitive thinking. In mathematics and formal logic, formalization means the reflection of a conceptual knowledge in the form of signs or by a formalized language. Formalization, i.e., the abstraction of notions from their content ensures knowledge systematization (separate elements of knowledge coordinate each other). Formalization plays an essential role in the development of scientific knowledge; intuitive notions are of little use for science (though they seem clear for a common mind). In scientific research, posing a problem (not to mention solving a problem) is often impossible without refining the structure of relevant notions. No doubt, true science proceeds from abstract thinking, successive reasoning of a researcher in the logical linguistic form (notions, judgements and conclusions).

Scientific judgements assist in establishing the connections among objects, phenomena, or their attributes. In scientific conclusions, a certain judgement is based on another; the existing conclusions lead to a new one. There are two primary types of conclusions – inductive (induction) and deductive (deduction) ones.

Induction is an inference from particular objects, phenomena to a common conclusion, from separate facts to their generalizations.

Deduction is an inference from the common to the particular, from general judgements to particular conclusions.

Idealization is the mental construction of beliefs about objects (nonexistent or unrealizable ones) whose preimages still exist in the real world. The process of idealization is remarkable for (1) abstracting from the properties and relations being inherent to real objects and (2) introducing (in the content of the resulting notions) attributes that could not in principle belong to their real preimages. The following notions are obtained by idealization: "a point," "a line" (in mathematics), "a material point," "a black body," "a perfect gas" (in physics).

The notions yielded by idealization are often said to possess idealized (or, simply, ideal) objects. Assume that idealization provides such notions about objects; in the sequel, one may operate them as really existing objects and construct abstract schemes of real processes for their better comprehension. In this sense, idealization is closely connected with modeling.

Analogy and *modeling*. Analogy is a mental operation such that knowledge (obtained by considering a certain object or model) is transferred to a less studied or less available (less visual) object called a prototype. Thus, a researcher opens up the feasibility of transferring data from a model to a corresponding prototype by analogy. This is the essence of a special theoretical method, *viz.*, modeling (building and analyzing models). The difference between analogy and modeling lies in the following; the former is a mental operation, while the latter can be considered as a mental operation or as a separate method-action.

Model is an auxiliary object that has been chosen or transformed for cognitive goals; a model gives new data about a corresponding primary object. The forms of modeling are diverse and depend on the models and their scope. According to the type of models, one distinguishes between physical modeling and sign (informational) simulation.

Physical modeling takes place for a model which reproduces certain geometric, dynamic or functional characteristics of the object being modeled (the prototype). In particular case, one has *simulation*, when the behavior of a prototype and that of a model are described by identical mathematical formulas, e.g., the same differential equations. Suppose that a model and a corresponding object possess the same physical nature; in this case, one speaks about *physical modeling*. In *sign simulation* the models are schemes, drawings, formulas, etc. An important type of such modeling is *mathematical modeling* (see below).

Modeling is always accompanied with other research methods (first of all, with experiments). Studying a certain phenomenon using its model represents a special type of experiments – a *model experiment*. In contrast to its common counterpart, a model experiment involves "an intermediate" during the process of cognizing, *viz.*, the model simultaneously being the means and object of experimental research (it replaces the prototype).

A special type of modeling is a *mental experiment*. During such an experiment, a researcher mentally creates ideal objects, correlates them with each other within a dynamic model; thus, a researcher simulates the dynamics and situations that could take place in a real experiment. Note that ideal models and objects assist in identifying (in the explicit form) the most essential interconnections and relations, as well as in playing mentally possible situations and rejecting unnecessary variants.

Moreover, modeling serves as the way of constructing something new (which has never existed in practice). Assume that an investigator has studied characteristic features of real processes and their tendencies; using a basic idea, he/she endeavors to find new combinations of the processes and tendencies, mentally reconstruct them (i.e., models the required behavior of the studied system). Similarly, any man or even animal creates his/her/its activity (demonstrates the property of active behavior) based on the "model of required future" formed initially (N. Bernstein). This process is accompanied with designing the models-hypotheses revealing the mechanisms of interconnections between the components of a studied object; later on, the described models-hypotheses are verified in practice. In such interpretation, modeling has recently become widespread in the humanities and social sciences (e.g., economics, pedagogics, etc.), when different authors propose different models of firms, production processes, educational systems, and so on.

In addition to the operations of logical thinking, the theoretical methods-operations include (may be, *de bene esse*) *imagination* as an mental process intended for creating new beliefs and images with its specific forms of fantasy (creating improbable, paradoxical images and notions) and *dreams* (creating the images of the desirable).

Theoretical methods (methods-mental actions). The philosophical, science-wide method of cognition is *dialectics*, i.e., the real logic of meaningful creative thinking which reflects the objective dialectics of the reality. Dialectics as a method of scientific cognition is based on ascending from the abstract to the concrete (G. Hegel), *viz.*, from general forms with poor meaning to the partitioned ones with rich meaning, to a system of notions enabling the comprehension of an object in its essential characteristics. In dialectics all problems acquire the historical character, and studying the development of an object makes up the strategic platform of cognition. Finally, in cognition process dialectics aims towards the revelation and solution of contradictions.

Dialectical laws (transition from quantitative changes to qualitative ones, the unity and struggle of opposites, etc.; the analysis of paired dialectical categories – the historical and the logical, a phenomenon and its essence, the general (the universal) and the single, and other laws) are inherent components of any competently organized scientific research.

Scientific theories confirmed by practice. In fact, any theory acts in the function of a method during construction of new theories (in the same or even another problem domain). Moreover, any theory acts in the function of a method defining the content and the sequence of the experimental activity performed by an investigator. Thus, the difference between a scientific theory as a form of scientific knowledge and as a method of cognition possesses the functional character. Being formed as a theoretical result of the previous research, a method acts as *terminus a quo* and as a prerequisite for further investigations.

Proof is a method, a theoretical (logical) action used to substantiate a certain idea based on other ideas. Any proof consists of three parts, notably, a thesis, reasons (arguments) and demonstration. Concerning the type of proofs, one should mention direct and indirect proofs. On the other hand, there exist inductive and deductive proofs (the classification based on the form of inferences). Let us note the following rules of a proof:

1   A thesis and arguments must be clear and precise.
2   A thesis must be identical during the whole proof.
3   A thesis must have no logical contradictions.
4   The reasons involved to substantiate a thesis must be true and indubitable; furthermore, they must not contradict each other and be a sufficient ground for a given thesis.
5   A proof must be complete.

In the whole set of methods of scientific cognition, a particular place is assigned to the *method of knowledge systems analysis*. To a certain measure, any scientific knowledge system appears independent of the corresponding problem domain being reflected. In addition, knowledge in such systems is expressed by a language, whose properties affect the attitude of the knowledge system to objects being studied. For instance, suppose that a certain well-developed psychological, sociological or pedagogical concept has been translated from English into German or French. Would this concept be unambiguously treated and understood in the Great Britain, Germany and France? Next, the application of a language as a notions' bearer in such systems presumes logical systematization and logical organization in using linguistic units to express knowledge. Finally, there is no knowledge system providing a comprehensive coverage for the content of a studied object. Merely a specific (historically concrete) part of the object's content can be described and explained within the framework of any knowledge system.

The method of knowledge systems analysis is relevant for numerous issues of empirical and theoretical research (choosing an initial theory, a hypothesis for solving a given problem; separating out empirical and theoretical knowledge, semiempirical and theoretical solutions of a scientific problem; substantiating the equivalence or priority of certain mathematical tools in different theories belonging to the same problem domain; studying the feasibility of extending the developed theories, concepts and principles to new problem domains; substantiating new opportunities of practical application of knowledge systems; simplifying and refining knowledge systems for learning and popularization; agreeing with other knowledge systems, to name a few issues).

Next, theoretical methods-actions comprise two methods of constructing scientific theories, namely:

- *the deductive method* (also known as *the axiomatic method*) is the way of constructing a scientific theory based on certain initial a priori true statements – *axioms* (*postulates*). Other statements of a given theory (*theorems*) are deduced from axioms by a logical proof. As a rule, theory construction by the axiomatic method is called deductive construction. All notions of a deductive theory (except a fixed number of initial notions, e.g., "a point," "a line," "a plane" in geometry) are introduced in the form of definitions involving other definitions (that have been introduced or obtained earlier). A classical example of a deductive theory is Euclidean geometry. The deductive method is used to construct theories in mathematics, mathematical logic, and theoretical physics;
- the second method has been assigned no name in scientific literature. However, it does exist, since the sciences (except the above-mentioned ones) develop theories according to a certain technique. For convenience, we will call it *the inductive-deductive method*; it consists in the following. First, one should accumulate an empirical basis. Subsequently, the latter serves to build theoretical generalizations – induction (this is probably done at several levels, e.g., empirical laws and theoretical laws); then the obtained generalizations can be extended to all objects and phenomena described by a given theory (deduction). See also Fig. 4.1 and Fig. 4.5. The inductive-deductive method is employed for most theories describing nature, society, and human beings (physics, chemistry, biology, geology, geography, psychology, pedagogics, etc.).

Finally, other theoretical methods of research (in the sense of methods as cognitive actions – the revelation and solution of contradictions, problem statement, hypothesis formation, and so on – right up to planning of scientific research) will be considered below according to the specifics of the temporary structure of research activity (designing the phases, stages and steps of scientific research).

Empirical methods (methods-operations).

The analysis of publications, documents and results of activity. Specific attention will be focused below on the issues of scientific literature survey. In fact, this is not just a method of research, but also an invariable procedural component of any scientific work.

Moreover, the actual material for any research includes various documentation (e.g., archival documents in historical investigations, internal documentation of enterprises, organizations and institutions in economic, sociological or pedagogical studies, etc.). Examining the results of activity plays an important role in pedagogics (especially, when analyzing the problems of professional training), psychology and sociology. Furthermore, in archaeology the analysis of the results of human activity discovered during excavations (the extants of tools, tableware, and dwellings) allows for reconstructing their way of life within the corresponding epoch.

Observation is, in principle, the most informative method of research. This is the only method displaying all sides of studied phenomena and processes being available to perception by an observer (we mean both direct observation and indirect observation using different devices).

Depending on the goals being pursued, one distinguishes between scientific and unscientific observations. The purposeful and organized perception of objects and phenomena of the reality, being related to solution of a specific problem, is called a *scientific observation*. Scientific observations imply obtaining some information for further theoretical understanding and interpretation, for confirmation or rejection of a certain hypothesis.

A scientific observation consists of the following procedures:

- defining the goal of observation (why or what for should one observe?);
- choosing an object, a process, a situation (what should one observe?);
- choosing the way and frequency of observation (how should one observe?);
- choosing the ways of detecting the observed object or phenomenon (how should one fix the obtained information?);
- processing and interpreting the obtained information (what should be the result of observation?).

The situations under observation are classified as:

- natural and artificial;
- the situations being controlled by an observer and being not;
- spontaneous and organized;
- standard and nonstandard;
- normal and extreme.

In addition, depending on the organization type, we single out open and hidden observations, field and laboratory observations. Depending on the character of fixing, it is possible to speak about ascertaining observations, estimating observations and hybrid observations. There exist direct and indirect (i.e., instrumental) observations (the classification basis lies in the way of obtaining the information). According to the scope of objects study, one may mention total and selective observations. Finally, there are permanent, periodic and single-time observations (the classification basis is the frequency of observation). A special case of observations consists in self-observations being common in psychology.

Observations are necessary for scientific cognition. Otherwise, scientists would be unable to accumulate initial information, to possess scientific facts and empirical data. Consequently, theoretical construction of knowledge would be impossible.

However, observation as a method of cognition has considerable shortcomings. Personal qualities of an investigator, his/her interests and psychological state may appreciably affect the results of an observation. To a larger extent, the objective results of an observation are subjected to distortion when a researcher strives to obtain a definite result (e.g., for confirming an initial hypothesis).

To guarantee the objective results of an observation, one should follow the requirements of *intersubjectivity*. Notably, during an observation the data must (if possible) be acquired and fixed by other observers.

Substituting a direct observation for its indirect counterpart (involving devices) enhances the opportunities of observations; nevertheless, such approach does not eliminate the subjectivity. Indeed, an indirect observation is assessed and interpreted by a subject, and the subjective impact of a researcher still takes place.

As a rule, observations are accompanied by another empirical method – measurements.

Measurement. Measurements are used all over the world, in any human activity. For instance, almost everyone controls and measures time daily. Here is a general definition of a measurement. *A measurement is a cognitive process which consists in comparing . . . a given quantity with its certain value accepted as a reference* (e.g., see [18, 33]).

In particular, a measurement forms an empirical method (method-operation) of scientific research.

It is possible to suggest a definite structure of a measurement with the following elements:

1 *a cognizing subject*, a person performing a measurement for specific goals of cognition;
2 *the means of measurement*, including different devices and tools designed by human beings, as well as items and processes created by the nature;
3 *the object of measurement*, i.e., a measured *quantity* or property being applicable for a comparison procedure;
4 *the mode of measurement* or *the method of measurement*, i.e., a set of practical actions and operations being performed by measuring devices, which includes definite logical and computational procedures;
5 *the result of measurement*, representing a denominate number in an accepted notation.

The epistemological grounds of a measurement method are closely connected with scientific interpretation of the ratio between qualitative and quantitative characteristics of a studied object (a phenomenon). This method serves for fixing only quantitative characteristics of an object; nevertheless, these characteristics are inseparably linked with the qualitative specificity of the object. The qualitative specificity enables identifying the qualitative characteristics for measurements. The unity of the qualitative and quantitative sides of a studied object means their relative independence and deep correlation. The relative independence of the quantitative characteristics allows for their study during measurements. Moreover, it allows for using the measurement results for analyzing the qualitative sides of an object.

The problem of *measurement accuracy* is also referred to the epistemological foundations of measurements as a method of empirical cognition. Measurement accuracy depends on the relationship between the objective and subjective factors in the measurement process.

Such objective factors comprise the following:

– the feasibility of identifying certain stable quantitative characteristics in a studied object; in many investigations (in particular, social and human phenomena and processes), the stated identification appears difficult or even impossible;
– the opportunities of measuring devices (the level of their perfection) and conditions during the measurement process. Sometimes, finding the exact value of a quantity is absolutely impossible (e.g., defining the trajectory of an electron in an atom).

The subjective factors of measurements include choosing the way of measurement, organizing the measurement process and the whole complex of cognitive abilities of a subject (e.g., consider pure experimentalists against researchers that can provide a competent interpretation to the results obtained).

In addition to direct measurements, in their experiments scientists often adopt the method of *indirect measurements*. In this case, a desired quantity is evaluated by direct measurements of other quantities being functionally related to the former. For instance, the measured values of mass and volume of a solid body enable estimating its density; next, the specific resistivity of a conductor can be computed by the measured values of its resistance, length and cross-sectional area, and so on. Indirect measurements are of crucial importance in situations when direct measurements are impossible in principle (e.g., the mass of any cosmic object is defined by mathematical computations proceeding from measurement data for other physical quantities).

Inquiry. This empirical method is used only in social sciences; there are oral and written inquiries.

Oral inquiry (conversation, interview). The essence of this method is evident from its name. During an inquiry an interviewer is in personal contact with an interviewee (i.e., the former can see the latter's response to a certain question). If necessary, an observer may ask additional questions to acquire supplementary data for uncovered issues.

Oral inquiries yield concrete results and assist in receiving comprehensive answers to complex questions of interest. However, one would better answer "ticklish" questions in written inquiries (providing a detailed and well-grounded treatment).

An oral answer requires less time and energy than a written one. But this method has a series of disadvantages. All interviewees are in nonidentical conditions (e.g., some of them may receive additional information through leading questions of an interviewer); a face expression or a certain gesture of an investigator has an impact on a respondent.

Written inquiry – questioning. It is based on a prepared *questionnaire*; the answers of respondents form the required empirical information.

The quality of empirical information obtained as the result of questioning depends on several factors: question formulation (it must be clear to respondents); qualification level, experience, honesty, psychological features of investigators; the situation and conditions of questioning; the emotional state of respondents; customs and traditions, beliefs, wordly situation; finally, the attitude towards questioning. Thus, in using such information, one should account for inevitable subjective distortions due to its specific individual interpretation in the minds of respondents. If the matter concerns very important issues, it is necessary to apply other methods (an observation, expert evaluation, the analysis of documents).

To acquire authentic data about a studied phenomenon or process, one does not have to question the whole contingent (the object of research can be large). The object of research exceeding several hundreds of respondents should involve sampling.

The method of expert evaluation. In fact, this is a type of questioning: the studied phenomena or processes are assessed by the most competent people whose opinions (supplementing and double-checking each other) enable rather efficient evaluation of the studied object. The applicability of this method depends on a series of conditions. First of all, through selection of *experts*, i.e., people being (a) aware of the assessed object and (b) able to provide an objective and open-minded evaluation.

The following are different methods of expert evaluation: the method of commissions, the brainstorm method, the Delphi approach, the method of heuristic forecasting – see the overview in [44].

Testing is an empirical method, a diagnostical procedure which consists in the usage of tests. A test generally represents a list of questions admitting brief and definite answers; alternatively, a test is a problem whose (unique!) solution requires a short time. Finally, a test may consist in short-term practical works of respondents, e.g., qualification tests in professional education, labor economics, etc. There exist printed-form tests, hardware tests (e.g., using a PC) and practical tests. One would also separate out tests for an individual and group application.

Thus, we have probably described all empirical methods-operations being available to the modern scientific community. We next consider empirical methods-actions that proceed from utilization of methods-operations and their combinations.

Empirical methods (methods-actions).

Empirical methods-actions should be divided into three classes. The first and second classes belong to studying the current state of an object.

The first class includes the methods of object analysis without its transformation (a researcher introduces no modifications in the object). More specifically, a researcher makes no essential changes in the object; indeed, according to the principle of complementarity (see above), an observer modifies the object anyway. Such methods are said to be *the methods of object tracking*. They comprise the method of tracking proper and its special cases – an investigation, monitoring, experience study and generalization.

Another class of methods is connected with an active transformation of a studied object by a researcher (*the methods of object transformation*). This class includes a *trial* and an *experiment*.

The third class of methods relates to examining the object's state with the course of time, *viz.*, in the past (retrospection) and in the future (*forecasting*).

Tracking appears the only empirical method-action in many sciences (e.g., in astronomy). As a matter of fact, today astronomers are unable to influence the cosmic objects being studied. Thus, it merely remains to track the state of such objects via methods-operations (observations and measurements). To a considerable degree, the same idea concerns geography and demography, where an investigator cannot change anything in the corresponding objects of research.

Moreover, tracking is widely used when the goal of research lies in studying the natural functioning of an object. For instance, consider investigations focused on certain characteristic properties of radioactive emission or the reliability of technical equipment during continuous service.

An investigation as a special case of an observation is studying an object with a certain level of "depth" and detailization depending on the goals posed by a researcher. An investigation is basically an initial study of an object, which serves for getting acquainted with its state, functions, structure, etc. As a rule, investigations are applied to organizational structures (enterprises, institutions, etc.) or to social groups – internal and external investigations.

External investigations include the investigation of sociocultural and economic situation in a region, the investigation of goods market, services market and labor market, the investigation of employment status, etc. *Internal investigations* are performed within an enterprise or institution (the investigation of the state of a manufacturing process, the investigation of employees).

Investigations involve the methods-operations of empirical research (observations, the analysis of documentation, oral and written inquiries, expert evaluation, etc.).

Any investigation is carried out according to a preset program with detailed planning of works, the corresponding tools (the preparation of questionnaires, tests, the lists of documents to-be-analyzed, and so on). Moreover, any investigation comprises certain assessment criteria for the phenomena and processes to-be-studied. Next, the following stages are necessary: data acquisition, results generalization, summing-up and preparation of reports. Each stage may require corrections of the investigation program (a researcher or a research group makes sure that the acquired data is insufficient for obtaining the desired results, or that the acquired data does not reflect the studied object, etc.).

Depending on the level of depth, detailization and systematization, investigations are classified as:

- pilot (exploring) investigations, being conducted for the preliminary (superficial) analysis of a studied object;
- specialized (partial) investigations, being conducted for the analysis of specific aspects and sides of a studied object;
- modular (complex) investigations, being conducted for the analysis of the whole blocks or complexes of issues programmed by a researcher based on the detailed preliminary study of an object, its structure, functions, etc.;
- systematic investigations, being conducted as full-fledged independent research based on the separation and formulation of its subject, goal, hypothesis, etc.; such

investigations imply integral consideration of an object and its backbone elements or factors.

What level should be chosen for organizing an investigation? A researcher or research group answers this question in a concrete situation, having in mind the posed goal or tasks of research.

Monitoring. This is permanent control, regular tracking of the object's state or parameters for studying the dynamics of running processes, forecasting certain events and preventing undesirable phenomena. For instance, take environmental monitoring, weather monitoring, etc.

Experience study and generalization. In any research, experience study and generalization (e.g., organizational experience, production experience, engineering experience, medical experience, pedagogical experience, etc.) serves for different goals. They are: assessing the current level of detailization of enterprises, organizations, institutions, assessing a technological process, identifying the shortcomings and "bottlenecks" in a certain practical activity, evaluating the efficiency of scientific recommendations, establishing new standards of activities as the result of creativeness of leading managers, specialists and collectives. The objects of research may include the following: (1) *mass experience* – for identifying the basic development trends in a certain sector of a national economy, (2) *negative experience* – for identifying typical drawbacks and "bottlenecks," and (3) *state-of-the-art experience* – for identifying, generalizing and making public (available to science and practice) new positive ideas.

State-of-the-art experience study and generalization are major factors of science development. Indeed, this method enables choosing relevant scientific problems, provides the base for studying the laws of process development in many sciences (in the first place, in the so-called technological sciences).

The criteria of state-of-the-art experience:

1 Novelty, being revealed at different levels (from introducing new statements in science to efficient application of well-known statements).
2 High efficiency. State-of-the-art experience must yield results having higher efficiency than the average results in a corresponding industry, within a group of similar objects, and so on.
3 Compliance with modern scientific achievements. Generally, attaining high results does not imply the correspondence between experience and scientific requirements.
4 Stability (preserving the efficiency of experience under varying conditions, ensuring high results during sufficiently large periods of time).
5 The feasibility of replication (using the accumulated experience by other people and organizations). State-of-the-art experience can be assimilated by other people and organizations. State-of-the-art experience might not be connected with personal peculiarities of its author.
6 Optimality (ensuring high results under relatively small consumption of resources, with no prejudice to solving other problems).

Experience study and generalization is implemented by many empirical methods-operations such as observations, inquiries, the analysis of publications and documents, etc.

A disadvantage of the method of object tracking and its versions (an investigation, monitoring, experience study and generalization as empirical methods-actions) lies in relatively passive role of a researcher. Notably, he/she may only study, track and generalize processes in the external environment (there is no opportunity to influence them). Again we emphasize this disadvantage is often caused by objective factors. On the other hand, *the methods of object transformation*, i.e., trials and experiments, do not suffer from such shortcoming.

Thus, the methods of object transformation include trials and experiments. The difference between them consists in the degree of arbitrariness of actions chosen by a researcher. That is, a trial is a nonrigorous research procedure, where a scientist introduces changes in an object at his/her own discretion (according to his/her own considerations of reasonability). In contrast, an experiment is an absolutely rigorous procedure – a researcher must adhere to the requirements of an experiment.

We have already mentioned that a *trial* is a method of introducing purposeful modifications in a studied object (with a certain degree of arbitrariness). For instance, a geologist chooses himself/herself what minerals to explore and what methods to use (e.g., to drill a well or to bore a pit). Similarly, an archaeologist or paleontologist decides the place and method of excavations. In pharmaceutics, new medications are the result of long-term research; just one of millions synthesized compounds becomes a medicinal agent. Another example is a trial in agriculture.

Trials as a research method are widely used in sciences connected with human activity (pedagogics, economics, etc.); they serve to develop and verify author's models (firms, educational institutions), as well as to develop and verify different original methods of authors. Alternatively, a trial textbook or trial medication is intended for practical tests.

In a certain sense, a trial resembles a mental experiment – they both address the issue "what if . . .?" The difference lies in that a situation is "played over" mentally in a mental experiment, while a real action is necessary in a trial.

Meanwhile, a trial is not a blind random search (it has nothing common with the trial-and-error technique).

A trial becomes a method of scientific research under the following conditions: being organized on the basis of scientifically obtained data according to a theoretically substantiated hypothesis; being accompanied by a deep analysis leading to conclusions and theoretical generalizations.

A trial involves all methods-operations of empirical research (observations, measurements, the analysis of documents, expert evaluation, etc.).

A trial is an intermediate between object tracking and an experiment.

This is a way of active interference in an object by a researcher. However, a trial demonstrates the efficiency or inefficiency of certain innovations in a general (overall) form. Which factors of implemented innovations ensure the highest or lowest effect? What is their interconnection? Unfortunately, a trial would leave these questions opened.

For a deeper analysis of the essence of a certain phenomenon, changes within it and their reasons, one often varies the conditions of such phenomena and the factors that influence on them. This is done within the scope of an experiment.

An experiment is a general empirical method of research (method-action), whose gist consists in the following. Phenomena and processes are studied in rigorously controlled conditions. The basic principle of any experiment is changing only a certain single factor in each research procedure (the rest factors are fixed and controlled). Suppose it is necessary to analyze the impact of another factor; then one should organize additional research procedure and vary this factor (provided that the rest factors are fixed).

During an experiment, a researcher *animo* modifies the pace of a process by introducing a new factor in it. A factor varied or introduced by an experimentalist is said to be an *experimental factor* or *independent variable*. Factors changed by the impact of an independent variable are called *dependent variables*.

One would find numerous classifications of experiments in scientific literature. First of all, depending on the nature of a studied object, a generally accepted classification includes physical experiments, chemical experiments, biological experiments, psychological experiments, and others. Next, according to the primary goal of an experiment, there exist *testing experiments* (the empirical verification of a certain hypothesis) and *search experiments* (the acquisition of necessary empirical data for constructing or refining a conjecture or idea). Based on the character and diversity of the means and conditions of an experiment (the ways of using these means), it is possible to separate out *direct experiments* (the means are directly used to study an object), *model experiments* (an object's model is used which replaces the latter), *field experiments* (in natural conditions, e.g., in space), and *laboratory experiments* (in artificial conditions).

Finally, we can consider qualitative and quantitative experiments (depending on different results of an experiment). As a rule, qualitative experiments are conducted for identifying the impact of certain factors on an analyzed process (without establishing a precise quantitative relationship between characteristic parameters). To guarantee exact values of essential parameters influencing the behavior of a studied object, one should organize a quantitative experiment.

According to the character of experiment's strategy, it is possible to distinguish among:

1   experiments implemented by the trial-and-error technique;
2   experiments based on a closed algorithm;
3   experiments involving the "black box" technique, leading from conclusions by the knowledge of a function to cognizing the object's structure;
4   experiments using the "open box," enabling the design of a sample with given functions (based on the knowledge of structure).

Recent years have been remarkable for wide adoption of experiments with computers as the means of cognition. They are of crucial importance when in real systems addressing direct experiments or experiments with material models is impossible. In some cases, computer experiments dramatically simplify research process (they serve to "reproduce" different situations by developing a model of a studied system).

Discussing an experiment as a method of cognition, we should emphasize another type of experimenting, which plays a major role in natural research. The matter concerns a *mental experiment*, when an investigator operates not a concrete (perceptional) material, but an ideal (model) image. All knowledge acquired during a mental experiment is subject to practical verification (in particular, using a real experiment). Therefore, the stated type of experimenting should be related to the methods of theoretical cognition (see above).

Moreover, some other types of experiments have to be classified as theoretical methods of scientific cognition (e.g., the so-called mathematical experiments and simulation experiments). The essence of a mathematical experiment lies in the following. Experiments are conducted not with an object (in contrast to a classical experiment), but with its model described within the framework of a corresponding branch of mathematics. A simulation experiment represents an ideal study by means of modeling of the object's behavior (instead of real experimenting). In other words, these types of experiments are modifications of a model experiment with idealized images.

Retrospection is a look in the past, a review of the past events. Retrospection research aims to study the state of an object and its development trends historically. Generally, retrospection research involves the technique of retrospection analysis.

Forecasting is a special scientific study of concrete development prospects of an object.

Thus, we have endeavored to describe research methods from very general positions. Naturally, each branch of scientific knowledge possesses well-established traditions in treating and applying research methods. For instance, the technique of frequency analysis in linguistics represents a method of object's tracking (a method-action), being implemented by methods-operations called the analysis of documents and measurement. Furthermore, it is possible to identify ascertaining experiments, learning experiments, check experiments and comparative experiments. Yet, the above-mentioned ones are the experiments (methods-actions) implemented by different methods-operations (an observation, a measurement, testing, etc.).

Chapter 4

Organization of scientific research

It has been emphasized above that a research project as a cycle of scientific activity includes three basic *phases*, notably, design phase, technological phase, and reflexive phase. Accordingly, we will consider cognition process in this logical structure, *viz.*, research design, research implementation (including results summing-up), assessment and self-assessment (reflexion) of the results.

Naturally, partitioning the research process into phases, stages and steps – see Table 4.1 (the temporary structure of research) – seems somewhat conditional.

Carrying out a research one often compares the intermediate results derived with the initial positions, with the draft version (project) of research. Thus, certain adjustments and corrections are continually introduced into the goals and course of research. In other words, assessment and reflexion run through the entire activity of an investigator. The reader should not be concerned with the fact that the reflexive phase appears as the last element in the discussed logical structure. Indeed, having completed a scientific work, a researcher generally initiates another (a new cycle of research) at a higher level – each work gradually enriches his/her experience.

Table 4.1 The phases, stages and steps of scientific research.

Phases	Stages	Steps
Design phase	Conceptual stage	Identifying contradictions Stating a problem Defining the goal of research Choosing criteria
	Modeling stage (hypothesis construction)	1. Forming a hypothesis; 2. Refining a hypothesis (concretizing).
	The stage of research planning	1. Decomposing (determining the tasks of research); 2. Analyzing the conditions (available resources); 3. Making up the program of research.
	The stage of technological preparations for research	
Technological phase	The stage of research implementation	Theoretical step Empirical step
	The stage of results summarization	1. Approving the results; 2. Formulating the results.
Reflexive phase		

To a large extent, the first phase (research design – from an idea to finite problems of research) is organized according to a common scheme: suggesting an idea – identifying a contradiction – stating a problem – defining the object and subject of research – formulating the goal of research – constructing a scientific hypothesis – choosing the tasks of research – research planning (scheduling of necessary activities). The logical structure of this phase is universally recognized. It has been formed on the basis of the centuries-old experience of research in all sectors of knowledge; moreover, this structure turns out optimal. Nevertheless, in each concrete case some deviations are possible (due to the specifics of a research object and tasks). For instance, historical investigations may have another structure.

The logic of the second (technological) phase of research can be very general. As a matter of fact, this logic almost fully depends on the *content* of a specific investigation (each research work is unique).

The logic of the last stage in the technological phase (approving the results, formulating the results) seems more univocal. It is common for most investigations and has been verified for years. This is the case for the logic of the third phase (reflexion, assessment and self-assessment of research results).

4.1 DESIGN OF SCIENTIFIC RESEARCH

The reader may pose natural questions. What does the term "research design" mean? What should be designed? The answer is easy – an investigator designs a *system of scientific knowledge* he/she actually wants to obtain. Indeed, in the beginning of this book we have outlined the following. The key aspects for a cycle of productive activity include: the developed model of a created system and the corresponding plan of implementation; implementing the system; estimating the implemented system and assessing the necessity of its correction or "launching" a new cycle. In the case of scientific research, the cited aspects can be expressed as follows: posing a scientific problem, constructing a scientific hypothesis as a cognitive model (these aspects are connected with design phase); during subsequent analysis, the above model-hypothesis is verified and assessed. Having been confirmed, the hypothesis forms a new system of scientific knowledge created by an investigator. Otherwise, the hypothesis is rejected, and one has to develop a new cognitive model – a new hypothesis (new hypotheses).

Design phase consists of several stages, namely, conceptual stage, modeling stage, the stage of research construction, and the stage of technological preparations for research. The names of these stages and steps are mostly imported from system analysis.

CONCEPTUAL STAGE IN DESIGN PHASE OF RESEARCH. The conceptual stage of design comprises the following steps: identifying a contradiction, stating a problem, defining the goal of research, choosing criteria (see Table 4.1).

Of course, when initiating a new scientific work, any researcher has an *intention*, i.e., a general idea of a project – what he/she actually wants to obtain. An intention comes into being under numerous conditions (practical requirements, the logic of science development, the preceding practical or research experience of an investigator, his/her personal preferences and interests). Probably, personal preferences and interests are the governing factors, since research is a creative activity (a point of great nicety). A lathe operator performs a routine activity (manufacturing the same component according to a drawing), a soldier implicitly obeys the orders of the commander. Contrariwise, an investigator must have definite freedom in choosing the direction, content and methods of research. Rich experience testifies that making a scientist work on a predefined goal (on a goal not chosen by himself/herself) appears pointless. An investigator independently chooses the subject and object of research, as well as thinks out the intention. Nevertheless, an investigator has to choose the type of research.

First, today there exists the following classification of research types (according to their "theory – practice" place):

- *fundamental research*, intended to design and develop theoretical concepts of science, its scientific status and history. The results of fundamental research may have no applications in practice;
- *applied research* mostly focuses on practical problems or theoretical issues related to practice. Generally, applied research represents a logical continuation of fundamental research, possessing an auxiliary (concretizing) character;
- *developments*. They serve for practical purposes.

Second, there exist different levels of research significance:

- the general level of significance – scientific works whose results exert an impact on the whole field of a certain science;
- the disciplinary level of significance – scientific works whose results contribute in the development of specific disciplines in a field of science;
- the problem-wise level of significance – scientific works whose results modify the existing scientific beliefs about a series of important problems within a certain discipline.
- the partial level of significance – scientific works whose results modify scientific beliefs about particular issues.

Suppose that an investigator has defined the intention of his/her research. Next, it is necessary to identify a contradiction.

The step of identifying a contradiction. A *contradiction* is the interaction between mutually exclusive (yet, mutually conditional and penetrating) opposites within a unified object and its states. It is well-known that identifying scientific contradictions makes up an important method of cognition. Scientific theories evolve as the result of discovering and solving contradictions in active theories and the practical activity of people.

The notion of a contradiction can be considered in two senses. First, when a certain entity (an expression, an idea) eliminates another one being inconsistent with it. Such interpretation of a contradiction (in a rigor sense) is often applicable to "exact" sciences (e.g., physics). The following are classical illustrations of contradictions that were formed at the end of the 1890s. Galileo's principle of relativity was at variance with Maxwell's equations in electrodynamics. The stated contradiction was solved within the framework of the special theory of relativity proposed by A. Einstein. In addition, remember the contradiction between the corpuscular theory of light and the wave theory of light that was settled by the development of quantum mechanics.

In social sciences and the humanities (still being by far less "exact"), a contradiction is treated in the second (less rigor) sense. Notably, a contradiction is an unconformity or incompatibility of two opposites, a disparity between the desired (e.g., according to the normative or theoretical viewpoint) and the actual (existing in real practice). However, the introduced definition of a contradiction underlies that opposites exist in a unified object.

A contradiction identified by a researcher may take place in the practice or theory of science; moreover, a series of contradictions may be found. Classical examples are contradictions in epistemologically strong sciences (physics, chemistry, etc.) – the results of an experiment exceed the limits of an existing theory (the development of scientific theories is discussed in [20, 21, 37]). Furthermore, partially examined problem domains (see examples in Table 4.2, Fig. 4.2, and Fig. 4.5) indicate of incomplete theories; in other words, this shows a contradiction – the disparity between the theory and the corresponding problem domain.

An identified contradiction allows an investigator to pose a problem of research.

The step of stating a problem. Choosing and substantiating a problem, as well as searching for its solutions play an essential role in the creative process of scientific cognition. *A scientific problem* is understood as a question that cannot be answered using the scientific knowledge accumulated by a society. Epistemologically a *problem* is a specific form of organization of knowledge whose subject lies not in the direct reality, but in the state of scientific knowledge regarding this reality. Imagine we are aware of knowing nothing about a certain object (e.g., its properties or the relations between its internal components). In this case, we have definite knowledge of a problem (the lack of knowledge).

For instance, we know that the nature of globe lightning has not been entirely studied. This knowledge of the lack of knowledge underlies the advancement of scientific problems.

A problem is a form of knowledge promoting the choice of direction in the organization of scientific research; indeed, it points to the unknown and stimulates cognizing the latter. A problem ensures purposeful mobilization of previous knowledge and organizing or obtaining new knowledge (as the result of research). A problem arises by fixing a real or forecasted *contradiction* by researchers; whether this contradiction will be solved or not affects the progress of scientific cognition and practice. Generally speaking, a problem is the reflection of a contradiction between knowledge and "knowledge about the lack of knowledge."

Science development is impossible without the requirement of goal-directedness. In scientific activity, *goal-directedness* is uniquely related to a problem. The reader knows – a problem points at the unknown and localizes it, thus performing the function of directing towards a goal. The stated goal-directedness appears explicit enough for defining the domain of the unknown; meanwhile, it is absolutely implicit for discussing the content of what should be cognized. During actualization of problems, an investigator often faces situations with a high level of uncertainty. This makes scientists address the structure of a studied problem and find criteria for (more or less visible) discrimination of actual and relevant problems.

Here an important role is assigned to the internal logic of the theory proper, since (if one has identified a problem being a foundation of the theory), solving the problem may generate a complete chain of corollaries. For instance, assume that the mankind has succeeded in describing all known types of interactions by a common physical theory (the so-called theory of everything or final theory). This would lead to theoretical forecasting and experimental discovery of numerous new physical phenomena and effects. Let us provide another example; the problems stated by D. Hilbert at the International Congress of Mathematicians (Paris, 1900) exerted a dominating influence on the development of mathematics in the 20th century. Still, many of the 23 Hilbert problems have not been settled.

The following steps accompany stating a problem: formulation, evaluation, justification and structuring of a problem.

1   *Problem formulation.* In problem formulation stating the questions is of crucial importance. Questions can be expressed in an explicit or implicit form (being rigorously defined or implied). First of all, problem formulation is the process of seeking for questions that gradually replace each other, thus approaching an investigator to the most adequate fixation of the unknown and to the ways of its transformation into the known. This is an essential aspect in problem formulation. However, problem statement is not completely reduced to this aspect. First, some scientific issues do not represent problems (they merely act as refining questions or unsolvable questions to date).

Second, a question itself is insufficient for problem formulation. One should also identify the foundations of a given question.

This is another procedure in the process of problem statement; the procedure serves to establish a contradiction which raises a problem question to-be-fixed.

Let us consider the following interesting example of fixing a contradiction underlying a scientific problem (see references in [29]). To be intelligent and skillful, one should have retentive memory and well-trained mind. Consequently, an inevitable contradiction takes place: investing much time in accumulation of knowledge means reserving less time for mind training, and vice versa. Thus, a certain optimal solution must exist. Many complications would be rejected if one found it.

For problem formulation it is important to construct an image, a "project" of the expected result of research (based on the forecast of research development and problem "background"). The "background" covers all circumstances of the problem (at the current and future steps) affecting the course and results of research.

2   *Problem evaluation.* Problem evaluation includes defining all conditions being relevant for its solution. Depending on the character of a given problem and opportunities of science, problem evaluation comprises choosing the methods of research, data sources, the staff of scientific employees and organizational forms required to solve the problem; in addition, we have to mention the sources of financing, the types of scientific discussion of the program and techniques of research, as well as the intermediate and final results, necessary scientific equipment and space required, possible cooperation partners in problem analysis, etc.

3   *Problem justification.* First of all, this is defining the meaningful, axiological (value) and genetic relations between a given problem and other ones (that have been solved earlier or being solved simultaneously with a given problem), as well as establishing connections to problems that would be feasible depending on solution of a given problem.

Second, problem justification is searching for weighty arguments in the favor of the necessity to solve a given problem (the scientific or practical relevance of expected results). This is the necessity to compare a given problem (or a given statement of a problem) with other ones in the context of choosing problems depending on their importance for practical purposes and internal logic of science.

Modern science often faces problems admitting several solution approaches. For instance, modern economics is remarkable for numerous models of firms, approaches to business organization, etc. Thus, one has to substantiate the choice of a solution or a model in specific conditions (e.g., as providing the maximal advantages). The more complicated is a problem, the greater number of various factors should be considered to justify its feasibility and plan its solution. The ability of a researcher to formulate and review the arguments used for substantiating the feasibility of a problem or accepting a suggested solution represents an important prerequisite of science development.

Assessing the relevance of a certain problem, one may observe overestimation of its actual value. Consequently, scientists gradually develop the so-called "defense reaction" (they incline to assess the actual value of any problem in smaller scales than the author of the corresponding research work revealing the problem). Such phenomenon seems rather natural in the scientific community. Indeed, science must be reasonably conservative to avoid going from one extreme to another on a certain subject. Meanwhile, this may cause underestimation of essential problems and lead to unjustified delays in the development of new scientific directions.

To reduce the subjectivity of problem estimation, it is important to raise objections against the problem (both by a researcher and his/her colleagues). One should dispute everything related to the essence of a problem, the conditions of its statement and possible consequences of its solution. Does the problem really exist? Is there any practical or scientific need in its solution? Is it possible to solve the problem at the modern stage of science development? Is the problem within the power of a given investigator or a research group? What is the possible value of the expected results?

A correct problem statement implies the battle of arguments "for" and "against." An accurate view of the problem's essence, the necessity of its solution, theoretical and practical value appears in the focus of opposite judgements.

4   *Problem structuring.* Problem structuring starts with its decomposition (known as *problem stratification*). Decomposition (see below) is identifying additional questions (subquestions) being vital for answering the primary question (solving a problem). One would hardly formulate all subquestions of a problem in the initial position. This often happens during research itself. Generally, it is difficult to find out all required for problem solution at the beginning. Hence, stratification (decomposition) is inherent to the whole process of problem solution. And so, initially one should choose and formulate all subquestions being necessary to begin research and reckon on an expected result.

According to V. Vernadsky, science always evolves in the following way. First, it decomposes a complex problem into simple ones; second, leaving the complex problem aside, science solves simple ones and afterwards addresses the complex problem.

During decomposition of a problem, it is important to localize the problem – to delimit the object of research (making it visible and feasible to an investigator or a research group) taking into account the conditions of research.

A scientist must be able to reject considering an interesting phenomenon or process that would complicate answering the central question (problem) of his/her research.

The separation (localization) of a problem is followed by ordering the complete set of questions (subquestions) of a problem according to the *logic of research*, i.e., making up a "network diagram" of subquestions solution.

Problem statement always involves the tools of a certain *scientific language*. The notions and structures of the language chosen to express a problem are far from being indifferent to its meaning. It happens that in the scientific community misunderstanding is connected not with complicacy of the problems proper, but with ambiguous usage of terminology.

Avoiding terminological confusion seems of crucial importance in the initial position of scientific research. During problem formulation and deployment one needs explicit definitions for all notions related to the problem. Moreover, ambiguities for problem setters can be successfully eliminated by stating a problem without technicalities. The benefits of a common language can be illustrated by a famous parody; a statement of a scientist is translated into a common language by a special robot. The scientist says, "Imagine four monocyclic units describing equidistant paths ..." The robot translates, "Imagine ... hem! ... four wheels."

Therefore, we have considered a specific form of organization of scientific knowledge, playing an important role in scientific research – a problem. Accordingly, the process of problem statement has been analyzed as a method of cognition.

Having posed the problem of research, an investigator defines the object and subject of research.

The object and subject of research. Epistemologically, *the object of research* opposes to a cognizing subject in his/her cognitive activity. In other words, the object of research is the ambient environment being faced by an investigator.

The subject of research is a side, an aspect, a viewpoint, a "projection" being used by a researcher to cognize a united object; thus, the researcher identifies the major (essential) attributes of the object. It happens that the same object represents the subject of different investigations or the whole scientific directions. For instance, the object called "educational process" may be studied by specialists in didactics, methodologists, psychologists, physiologists, hygienists, etc. However, they would definitely focus on various subjects of research. Moreover, the subject of a certain research may act as the object of another one (a particular research). Consider the object called "the quality of life"; it can be analyzed within the framework of medical science, economics, sociology, etc. An important aspect of this object ("population health") is the subject of research for medicine. On the other hand, the above-mentioned aspect makes up the object of research for a branch of medicine – public health organization.

Let us discuss the correlation between the object and subject of research in a greater detail (see references in [29]).

The subject of research emerges as the result of definite cognitive operations with the object of research. The subject of research represents the totality of properties – relationships and laws being studied by a given science and expressed in specific logical and sign forms. This feature distinguishes the subject of research from the object of research; the latter exists regardless of a cognizing subject (in nature, in a human being or a society).

Moreover, the difference between the subject and object of research consists in the following. The same object of research may be examined by many sciences (each science separates a special subject of research). For instance, cosmic objects are studied by astronomy, astrophysics, astrobotany, and so on. Society as the object of research is analyzed by history, political economy, philosophy, demography, etc. All these sciences possess nonidentical subjects of research.

The subject and object of research vary in their structures. Notably, the structure of the object of research is the interaction among the basic components of the object. Such interaction among the basic components generates different properties, relations of the object and the laws of its development. To a certain degree, the structure of the subject of research is determined by the structure of the object of research; this determination appears nonrigorous though. The structure of the subject of research is relatively independent. It comprises the following basic components: first, the history of science development concerning the studied object of research; second, the essential properties and development laws of the object (that have been expressed in definite logical forms as the result of cognition process); third, the logical framework and methods being employed to form the subject of research.

In many respects, the structure of a subject of research depends on the level of cognition which serves to form the subject. At the empirical level, the subject of research is directly connected to the object; here all cognitive operations are performed by means of many methods such as observations, measurements, etc. These methods assist in fixing, comparing and classifying all empirical data about a studied object. According to the stated data, the subject of empirical research includes the following elements: first, all fixed facts concerning the behavior of a studied object; second, all measurement data for different properties and relations of a studied object; third, signs and sign forms involved to register empirical data; fourth, all statistical data regarding internal changes and development (the emergence or extinction) of such properties and relations of a studied object, being discovered during empirical research.

Hence, at the empirical level of cognition the subject of research mismatches the object of research. The subject of research expresses merely phenomena, their properties and relations that have been successfully fixed, classified and expressed using sign forms. Thus, everything points to the fact that mediation of the subject of research takes place as early as at the empirical level. The relation between the subject and object of research is mediated by the statistical data about studied phenomena, as well as by the logical tools of their expression and preceding knowledge (providing the base to perform all empirical cognitive operations).

Further mediation of the subject of research continues at the theoretical level. The subject gradually strays, abstracts away from the object of research. The theoretical level is to analyze the empirical data. The latter assists in revealing the essence of studied phenomena, their properties and relations, as well as in establishing the development laws of studied objects, formulating scientific hypotheses and theories, and making scientific forecasts. At the theoretical level, cognitive operations condition the specifics of the subject of research. Consequently, the subject of research covers and expresses the essential and deepest characteristics and properties of a studied object. The subject of research is now related not to specific phenomena, but to their development laws.

Development laws, scientific hypotheses and theories constitute the basic features of the subject of research at this level.

The notions "the object of research" and "the subject or research" fulfill different functions in the cognition process. The former notion expresses and fixes the objective existence of studied phenomena, their properties, relations and development laws. This notion aims investigators towards most comprehensive reflection of the essential objective sides of a studied object in different forms. The more accurate and complete is such reflection of objective sides in knowledge, the deeper is the scientific content of the knowledge. "The object of research" acts as an initial notion to interpret the content of our knowledge.

First of all, the notion "the subject of research" defines the limits used to study a certain object. This notion expresses and fixes the properties, relations and development laws of a studied object, which have already been included in scientific knowledge and expressed in definite logical forms. Suppose that a certain science goes beyond its subject. This means either incompetent interference of a given science in the subject of other sciences, or splitting a given science into new branches. Later on, new branches may have a separate subject of research.

In this context, one may provide several positive examples such as physical chemistry and molecular biology arising at the junction of other sciences that have reached a certain level of development. Yet, there exist negative examples (ungrounded analogies and/or extension of the subject of research). Note that the representatives of both epistemologically weak sciences and epistemologically strong sciences are given to it. Indeed, an investigator performs a pedagogic experiment in a single educational institution and claims that the derived results hold true for any educational institution (there is a direct evidence of ungrounded extension of the subject of research, being accompanied by ungrounded transfer of results from a certain subject to the other). Moreover, one often deals with scientific works, where a mathematician employs a perfectly mastered framework in a new problem domain without a detailed analysis of the latter's specifics (this points to the usage of ungrounded analogies). In both cases mentioned, the validity of results is subject to natural doubts (the criteria of assessing scientific theories are discussed below).

The subject of research serves to formulate cognitive problems of a certain science (in a concentrated form). Furthermore, it serves to choose the primary directions of scientific research and to define the feasibility of the corresponding cognitive problems by the means and methods of a given science. **To uniquely characterize the activity of a certain investigator, it suffices to specify the subject of his/her research and the methods being employed** (see Fig. 4.2). The periods of intensive development of a certain science take place when its subject gets extended or new methods appear. For instance, consider astronomy; initially, it studied stellar sky by the observation method. Gradually, the subject of astronomy was extended (by embracing the range of problems connected with the origin and development laws of the Universe) and it turned into astrophysics. Rapid development of the latter caused the emergence of new theories (and their experimental verification – e.g., recall the discovery of the Universe expansion in the 1920s) or to the design of new experimental devices (e.g., radio telescopes).

However, a somewhat paradoxical negative example demonstrating the absence of a well-defined subject of research is provided by the scientific direction known as *operations research.* This field of applied mathematics analyzes solving the problems of operations modeling (modeling of purposeful actions), i.e., phenomena in economics, production area, social systems, and so on [15, 43]. Numerous publications are dedicated to this scientific direction; however and unfortunately, none of researchers has managed to correctly define the subject of this "science." As a matter of fact, it boils down to the groups of separate problems being solvable today. Such a situation is natural for many scientific directions, whose limits are determined not by the subject of research (a well-defined problem domain), but by the set of derived theoretical results (sometimes, disconnected ones). Moreover, presently one may find textbooks for students describing "new" educational courses without the definition of their subject of research (being committed to scientific ethics, here we abstain from providing concrete examples).

Therefore, the dialectical correlation between the object and subject of research is of paramount importance in the process of scientific cognition. It enables scientific interpretation for the content of knowledge being formed during research. In addition, it enables rigorous definition of the limits to-be-observed by a given science in using its means and methods to study objective phenomena, their properties, relations and development laws.

Apparently, competent definition of the object and subject of research is not a trivial problem. It gets more sophisticated in the case of large-scale generalizing investigations being the fruit of many years' research work of a single scientist (a series of separate studies) or a group of scientists. Prior to determining the object and subject of a generalizing research, one should clearly specify the corresponding *problem domain* – the whole set of phenomena described by a given theory.

An investigator undertaking such a generalizing research obtains numerous (diverse and many-sided) results; it would be difficult to integrate them.

Thus, an investigator has to search for a problem domain (a topic or a concept) being able to integrate all accumulated results (or, at least, most of them). It happens that some results appear impossible to combine within a common framework. Hence, they should be rejected. At the same time, sometimes additional results are required, and research should be continued. In this context, let us provide the following analogy from set theory (see Fig. 4.1 with the Euler-Venn diagram). Suppose there are separate results – sets 1, 2, 3, 4, etc. (see Fig. 4.1a). The sets may be partially "overlapping." The problem lies in finding a set (referred to as a generalizing set, see Fig. 4.1b) that would include all or, at least, most of the separate sets. Again we note that sometimes separate results not relating to a specific problem domain have to be "rejected" (sets 8–9 in Fig. 4.1b).

As a rule, such a uniting problem domain is successfully identified. Let us endeavor to describe an approximate "algorithm" of this search. Pose the following question. What is the origin of new results providing a possible base for a generalizing research? Imagine three conditioned planes (see Fig. 4.2), *viz.*, the plane of problem domains, the plane of methods and means of research (the so-called "technology" of research), and the plane of results.

New results can be obtained by:

1   studying a new (unexplored) problem domain – see Fig. 4.2a (here "novelty" is marked by shading);

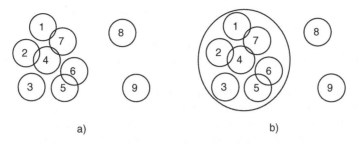

Figure 4.1 The Euler-Venn diagrams. Finding a "unitizing" set.

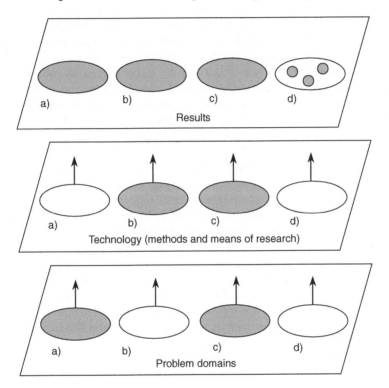

Figure 4.2 The ways of obtaining new scientific results.

2   applying new *technologies* (methods or means of research) to an explored problem domain – see Fig. 4.2b; for instance, to study a certain problem domain, one adopts a new *research approach* or theory from another field of scientific knowledge (recall a theory may act as the method of research); alternatively, it is possible to use a specific mathematical tool (as the means of research) that have not been employed in a given problem domain before; another example is involving new material means (e.g., new devices or linguistic tools), and so on;

3   applying new technologies to study a new problem domain – see Fig. 4.2c.

Interestingly, in some fields of science researchers are traditionally divided into two categories. The first one is called "screwmen" (in a certain sense, such researchers "unscrew," i.e., study new problem domains). The second one is known as "spannermen" (such researchers use new technologies of cognition, i.e., "fit spanners for unscrewing"). An investigator has to choose the proper category depending on his/her idea and derived results.

Clearly, another way of obtaining new results (see Fig. 4.2d) is impossible in principle; indeed, one would not generate new results or make generalizations by considering the same problem domain or using well-known technologies.

The following law takes place: the wider a problem domain is, the more difficult is obtaining new common scientific results for it. This phenomenon vividly shows itself in *mathematics*. Notably, any formal assertion (e.g., a theory) consists of two parts – suppositions ("Let ...") and inferences ("Then ..."). The stronger are suppositions (conditions, constraints), the simpler is the proof and the deeper are the results.

Recall sciences are decomposed into epistemologically strong and weak ones (see Chapter 1). Accordingly, the stated law can be reformulated as follows. Epistemologically weak sciences introduce the minimal constraints (or no constraints at all) and obtain the fuzziest results. Contrariwise, epistemologically strong sciences impose many limiting conditions, involve scientific languages, but yield more precise (and well-grounded) results. However, the field of their application appears rather narrowed (i.e., clearly bounded by these conditions).

Any suppositions (constraints) confine the domain of applicability (*validity*) of the corresponding results. For instance, in control of socioeconomic systems, mathematics (operations research, game theory, etc.) suggests efficient solutions; but the domain of applicability (adequacy) is appreciably limited by the explicit suppositions made to construct the corresponding models. On the other hand, social sciences and the humanities (also treating control problems in socioeconomic systems) introduce almost no suppositions and propose "universal remedies" (i.e., their domain of applicability is rather wide). But the efficiency of such "remedies" often coincides with that of *sensus communis* or the so-called best practices (the generalization of a positive practical experience). Without appropriate investigations one would hardly guarantee that a management decision (proved its efficiency in a certain situation) preserves the efficiency in another (though, a very "close") situation.

Thus, it is possible to draw different sciences using the coordinate axes "The Domain of Validity" and "The Domain of Applicability"; by analogy to the Heisenberg uncertainty principle), one can *de bene esse* formulate the following *principle of uncertainty*. **The current level of science development is characterized by certain mutual constraints imposed on results "validity" and results applicability**, see Fig. 4.3. That is, the "product" of the domains of results applicability and validity does not exceed a constant (increasing the value of a "multiplicand" reduces the value of another "multiplicand").

The aforesaid does not imply that the development is impossible in principle – each specific research represents advancement towards the growth of "validity" (generality) and/or towards extending the domain of applicability (adequacy). As a matter of fact, the history of science development testifies to the shifts of the curve shown in Fig. 4.3 to the right and to the top (i.e., it illustrates the growth of the above-mentioned constant)!

An alternative explanation has the right to exist, as well. "Weakening" of sciences takes place as soon as the object of research gets complicated. Consequently, all

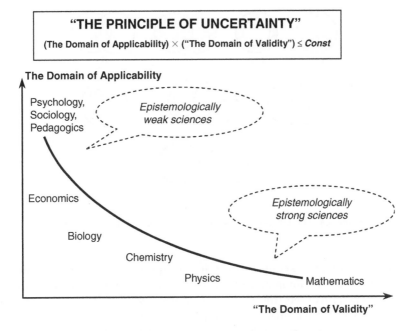

Figure 4.3 Illustrating "the principle of uncertainty".

epistemologically strong sciences can be called "simple," while the epistemologically weak ones can be referred to as "complex" sciences (based on the complexity their object of research). An imaginary "boundary" between them is biology (living systems). Analyzing separate systems of an organism (anatomy, physiology, etc.) still has a propensity for epistemologically strong sciences (the empirics is confirmed by reproducible experiments and is grounded by "simpler" sciences – biophysics, biochemistry, etc.). Thus, it may provide a base for formal constructions (similarly to physics and chemistry). Next, for living systems the experiments in their classical interpretation (reproducibility and so on) become difficult. Moreover, they are almost impossible for human beings and social systems.

Let us get back to a detailed description of different definitions of the problem domain of generalizing research. It is possible to provide a definite topology here.

Again we involve the analogy from set theory – the *Euler-Venn diagrams* (see Fig. 4.4 with new problem domains being shaded).

The following cases are then possible.

Case (a). A separate set (the analog is a new problem domain). This case, i.e., the appearance of an absolutely new problem domain, occurs infrequently (as a rule, due to his/her education a researcher has blinkers on when it comes to investigations). However, exactly this case may lead to a scientific breakthrough (the appearance of new scientific directions).

Case (b). The inclusion of a set in another set (the analog is extending a problem domain). This case seems to be the most typical in evolutionary development [20] of a theory or a scientific school. A problem domain is extended by enlarging the subject of research, by generalizing the obtained results, etc. For instance, in mathematics this

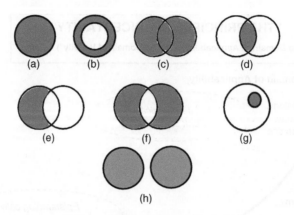

Figure 4.4 The Euler-Venn diagrams. The basic operations with sets.

case corresponds to a weakening of introduced suppositions with preservation of the derived results or to obtaining new (more general) results under existing suppositions.

Case (c). The sum of sets (the analog is forming a problem domain based on shared elements of two problem domains). The typical example lies in *generalization* yielding a theory to unite two theories with intersecting problem domains. This case (similarly to case (b)) is inherent in evolutionary development; still, it may reflect revolutionary aspects of development of a certain theory (all depends on the size of problem domains). An example from physics is the electroweak theory proposed in the 1960s by C. Yang and R. Mills; this theory describes electromagnetic interaction and weak interaction via a common framework.

The cases (a)–(c) correspond to the extension of a problem domain; on the contrary, cases (d)–(g) deal with its narrowing. Since the subject of research gets narrowed, obtaining new scientific results requires new approaches, methods and means of research.

Case (d). The intersection of sets (the analog is forming a problem domain based on shared elements of two problem domains). This case corresponds to obtaining deeper results as opposed to the ones derived in appropriate problem domains (by narrowing a problem domain); note this seems somewhat exotic. Alternatively, case (d) corresponds to the transfer of results (generally, methods of research – see Fig. 4.2c) from a problem domain to another, or to practical interpretations of the results (derived in a problem domain) in terms of another problem domain. As an example, recall the beginning of the 20th century when the framework of differential equations was successfully applied to ecosystems description (the dynamics of interaction among biological populations, the competence among biological species, etc.). Previously differential equations were used mostly in physics and engineering.

Case (e). The difference of sets (the analog is forming a problem domain by excluding elements of a first problem domain from a second problem domain).

Case (f). The symmetrical difference of sets (the analog is forming a problem domain based on nonintersecting elements of two problem domains). The cases (e)–(f) correspond to restricting a problem domain, when the subject of research represents, e.g., objects possessing only a given property and not possessing another property (case (e)) or possessing a single property from two given ones (case (f)). For instance, a researcher analyzes the process of adaptation of an individual after retiring on a pension (in case (e)), the initial sets are the set of pensioners and the set of working people; the shaded set indicates non-working pensioners). In case (f), an example is provided by a biomedical research of the comparative efficiency of two different medications against a disease. The case of therapy using both medications is excluded.

Case (g). Set narrowing (the analog is extracting a set of elements with identical properties from a problem domain to form a new problem domain). Such situation is common in epistemologically strong sciences, when existing results are strengthened by imposing some limiting suppositions (see an illustration for the principle of uncertainty in Fig. 4.3). For instance, there exist numerical solution methods for algebraic equations of arbitrary order. At the same time, algebraic equations whose order does not exceed three (a narrower problem domain) admit analytical solutions.

Case (h). Two nonintersecting sets. Evidently, this case includes *comparative research* (e.g., the comparative research of legal systems in two countries).

We have studied the ways of forming problem domains, corresponding (by analogy) to all basic operations with sets. Thus, one may conjecture that this set of operations (supplemented by their feasible combinations) covers all possible ways of defining problem domains. Therefore, conducting a generalizing research in a given problem domain, one determines the object and subject of research.

The topic of research. A reader may pose a natural question as follows. Why has the *topic of research* still not been mentioned? Indeed, the topic or research should be of primary importance, preceding its idea, contradiction, problem, etc. No doubt, a rough formulation of the topic of research is provided from the very beginning of investigations. However, generally a scientist provides a concrete statement for the topic or research after defining the subject of research. In the overwhelming majority of cases, the topic of research points to the subject or/and the methods of research; moreover, key words used to formulate the topic of research often indicate the subject of research.

In addition to the object of research, the content and direction of research depends on research approaches. The category "research approach" acts in two interpretations.

According to the first interpretation, an *approach* represents a certain initial *principle*, a zero position, a basic statement or belief (e.g., holistic approach, complex approach or functional approach in engineering). For instance, one may face informational approach or cybernetic approach). In this sense, we should emphasize system approach, complex approach and synergetic approach as the most popular ones.

In its second interpretation, *research approach* is treated as the direction of studying the object of research. Such approaches possess general scientific character. They appear applicable to investigations in almost any field of science and are classified using pair dialectic categories reflecting polar sides and directions of cognition process (content against forms, the historical against the logical, quality against quantity, phenomena against matter, and so on).

Substantial approach and formal approach. Evidently, a *substantial approach* requires addressing the content of studied phenomena and processes, identifying the totality of their elements and interactions defining the basic type or nature of these phenomena. Furthermore, the above approach requires referring to facts and data (accumulated during observations and by experience) and deriving theoretical inferences based on them (via abstractions, analysis, and synthesis).

On the other hand, a *formal approach* stipulates extracting merely stable or relatively invariant features from examined processes and phenomena; they are considered "in the pure sense," i.e., independently of the whole process or phenomenon. Note here the term "formal" bears no negative meaning (many word combinations such as "knowledge formalism," "the formal attitude" of a bureaucrat are negative). Formal approach (also known as formalized approach) allows revealing stable relations among elements of a process or phenomenon.

To grasp the difference between substantial and formal approaches, let us provide the following example. Consider the problem of students with poor grades. For instance, establishing the social reasons of this problem makes it necessary to apply a substantial approach. Yet, identifying the statistical laws of the corresponding dynamics by years (or the distribution by different regions) can be performed within a formal approach.

The use of any mathematical tools or models of phenomena and processes, the application of any symbolic or formula languages is actually the implementation of formal approach.

Of course, substantial and formal approaches are interrelated and interdependent. As a rule, a formal consideration of a subject must be preceded by its substantial analysis. At the same time, formalization (translating a substantial knowledge into an artificial language) is supplemented by the inverse process – *interpretation* (a substantial explanation of formal results).

We underline that formal approach is not in the least connected with quantitative approach (see below). For instance, in a series of investigations researchers involve the elements of topology and graph theory (far from often these fields of mathematics operate the categories of quantities and numbers).

Logical and historical approaches. The dialectic principle of historicism implies the unity of logical and historical ways of cognition while studying developing objects. The logical approach reproduces an analyzed object in the form of its theory, whereas the historical one reproduces the object in the form of its history. Naturally, the both approaches are mutually complementary.

Logical approach means consideration of a phenomenon or process in the present point of its development; in this case, abstract-theoretical constructions dominate the research.

Historical approach lies in analyzing the concrete-historical genesis (origin) and development of an object, as well as in studying and reflecting mainly genetic relationships of a developing object. And so, concrete historical facts prevail in the research.

One should keep in mind the necessity of uniting historical and logical approaches (as supplementary and interweaving ones).

It may be reasonable to adopt *logical-historic approach*, when revealing a studied problem combines historical approach (the historical development of phenomena, processes and scientific ideas or theories) and logical approach (the modern state of phenomena, processes and scientific ideas or theories, including their correlation). However, in a logical-historical approach the logical aspect has primary importance.

An alternative consists in *historical-logical approach*; here one would observe the dominance of the historical aspect.

Qualitative and quantitative approaches. *Qualitative approach* aims to explore the totality of attributes, properties and specific features of a studied phenomenon or process (determining its uniqueness, singularity and belonging to a certain class of similar phenomena or processes). *Quantitative approach* aims to explore the characteristics of various phenomena or processes by the development level or intensity of properties being inherent to them and expressed in quantitative terms (using quantities and numbers).

Evaluating the quantitative characteristics of subjects, phenomena or processes starts from identifying their common properties being intrinsic either to homogeneous or heterogeneous phenomena or processes (in the sense of their nature). As a matter of fact, such identification "obliterates" qualitative differences between the latter and results in some unity (thus, enabling *measurements*). For instance, a person is a unique individual, and introducing certain quantitative characteristics to assess personalities of different people seems impossible. Nevertheless, one can compare people using common indicators (e.g., height, weight, etc.) – properties observed for everybody.

Discussing various classifications of research approaches (based on pair categories of dialectics), we separate out **phenomenological approach** and **essential approach**. The former aims to describe externally observed (generally, variable) characteristics of a certain studied phenomenon or process. The latter aims to identify their internal and stable sides, mechanisms and driving forces.

Phenomenological approach appears legitimate during definite stages of science development. For instance, C. Linnaeus (on the one part) and C. Darwin (on the other part) succeeded to create a classification of biological species and evolution theory, respectively, exclusively due to generalization of bulky real (phenomenological) material accumulated by biology. Another example concerns Kepler's laws of planetary motion; they were obtained by generalizing numerous observations and measurements performed by Danish astronomer T. Brahe.

Finally, in this sequence of research approaches let us note **single and general (generalized) approaches.** Clearly, a *single approach serves to* study separate phenomena or processes, whereas a *general approach* serves to find their common relations, laws and typological features.

The above-mentioned classifications of research approaches by pair categories of dialectics are independent. Hence, each research can be characterized by their specific set. Moreover, different problems within the same research can be solved by different sets of approaches.

Yet, the category "research approach," as well as its role and place in the structure of methological knowledge have been insufficiently analyzed. The existing ambiguities in this issue can be illustrated by a simple example. Above we have classified research approaches (in the second interpretation) by five pairs of dialectic categories. Consequently, the same subject of research may have $2^5 = 32$ different research approaches. Notably, given a certain subject of research, a scientist may carry out 32 totally different types of investigations! Moreover, the number of feasible approaches in the first interpretation (system approach, personal approach, synergetic approach, etc.) is infinite!

In aesthetics, art criticism (the study of art), the theory of literature (see [29]), the notion of a *method* of an art or literary work represents a certain analog of research approach (e.g., classicism method, romantism method, realism method, etc.). In architecture we have the notion of a *style* (e.g., classical style, Empire style, art nouveau, etc.). Similarly, in science research approaches play the role of (special!) methods. In Section 2.2 we have divided research methods into two levels, *viz.*, methods-operations and methods-actions. In this context, research approaches form the third level – they appear *supermethods*.

The step of defining the goal of research. Proceeding from the object and subject of research (and chosen approaches), a scientist determines the goal of research. Generally speaking, the *goal of research* means what should be-achieved at the end of research.

Naturally, it seems easy and logically correct (anyway, formally correct) to formulate the goal of research in the short statement as follows. "The goal of research is to solve the posed problem of research." Of course, the problem of research must be rigorously posed. However, in this case a researcher has the courage to claim the following. He/she has completely settled the problem and other investigators would definitely contribute nothing to it. No doubt, D. Mendeleev exhausted the problem of chemical elements classification by establishing the periodic law. Alternatively, A. Einstein solved the problem of correspondence between mechanics laws and electrodynamics laws by suggesting special theory of relativity. Still, claiming that a researcher has completely explored a problem appears risky or even venturesome. Anyway, we assume that (by the end of research) one should completely solve the posed problem (within the limits specified by the subject, object and goal of research – see below).

Note that many research works in the field of social sciences (especially, Candidate's thesis) include incorrect statements of the goal of research. This happens when, defining the expected scientific result (a new knowledge as the basic outcome of any research), the authors set themselves the task of achieving practical goals. For instance, such goals as "improving the process of ...," "increasing the efficiency of ..." and others could not be the goals of scientific research. Later on, under certain conditions (e.g., implementation), scientific results may represent the base for "increasing the efficiency"; nevertheless, one may not pose such goals of research. Even such formulations as "the goal of research is to develop scientifically substantiated recommendations ..." act merely as supplementary or auxiliary goal of research (or rather as a certain problem of research increasing the practical relevance of research).

The step of choosing assessment criteria for validity of research results. Suppose that an investigator has defined the goal of research (i.e., the results to-be-obtained and their structure are clear). Consequently, the investigator starts planning and choosing assessment criteria for validity of the expected research results. This issue turns out the most complicated and critical for any research (what criteria should be used to assess innovations or theories?). As a matter of fact, *criteria* are of crucial importance in any activity. Inaccurate choice of criteria may lead to collapses of social institutions and economic systems.

Therefore, starting a research work, one has to seriously approach the choice of assessment criteria for validity of research results. We underline that the assessment criteria for validity of theoretical research results are well-defined (they have been matured over many years' experience in investigations). Yet, the assessment criteria for validity of empirical research results are invidiual for a concrete research work, since they totally depend on its content. Nonetheless, there exist some common recommendations regarding their choice. We discuss them below.

The assessment criteria for validity of theoretical research results. The result of a theoretical research (a *theory*, a *concept* or some theoretical constructions) must satisfy the following universal criteria (principles) formulated for any fields of scientific knowledge (also, see references in [29]):

1 single-subjectedness;
2 completeness;
3 consistency;
4 interpretability;
5 verifiability;
6 validity.

Single-subjectedness as an attribute of a scientific theory means that the whole set of notions and assertions of the theory must belong to the same problem domain. The attribute of single-subjectedness does not disclaim the existence of several theories describing the same phenomena or processes (this fits the principle of complementarity – see above).

Completeness as an attribute of a scientific theory means that the theory must cover all phenomena and processes within its problem domain.

Consistency as an attribute of a scientific theory means that all postulates, ideas, principles, models, conditions and other structural elements of this theory must not logically contradict each other.[1] Indeed, identifying and solving contradictions in scientific theories motivate their further perfection and development, as well as construction of new theories.

Interpretability as an attribute of a scientific theory (first of all, a formal theory) means that the theory must possess an empirical content and provide for practical interpretation of formal results. There is no theory without a practical interpretation (otherwise, it simply represents a set of signs and formulas). An exception lies in mathematics; for instance, Lobachesky's geometry initially was a pure abstraction without a practical interpretation.

The attribute of *verifiability* of a scientific theory characterizes it in the sense of substantial *truth* and ability for development and perfection. Verifiability acts as setting the correspondence between the content of theoretical statements and the properties of real objects. In many cases, the only option is to verify such correspondence in order to set it.

The attribute of *validity* of a scientific theory means that the truth of its basic statements has been reliably established. Thus, a scientific theory differs from a scientific hypothesis (where truth is established to a degree of reliability).

Unfortunately, many (not to say most) researchers in the fields of social sciences at various levels of scientific hierarchy even do not suspect of the existence of such attributes and requirements applied to a scientific theory or concept. As a rule, in

---

[1]Of course, the completeness and consistency of any theory are relative. For instance, consider mathematics; the well-known Hedel theorems state that any sufficiently complex theoretical system is incomplete (on the one part) and its consistency could not be fully shown within the framework of this system (on the other part).

theory publications authors introduce numerous principles, conditions, technologies, etc. in the form of arbitrary "enumerables" such as purposefulness, fundamentality, technological effectiveness, dynamism, openness and so on. Meanwhile, most speakers at a scientific conference would get into a mess if someone makes the following simple request: "Please, prove the completeness (or consistency) of your concept."

Naturally, the above-mentioned attributes – criteria of a scientific theory or concept – appear elementary. They assist in preliminary assessing the results of a theoretical research (upon its completion). The ultimate assessment criterion for validity of a scientific theory consists in its practical implementation. "There is nothing practical than a good theory." However, this criterion requires (generally, much) time.

The assessment criteria for validity of empirical research results. The assessment criteria for validity of theoretical research results must have the following attributes:

1   The criteria must be *objective* (as much as possible in a given problem domain), enable explicit assessing the studied attribute without disputable appraisals by different people.
2   The criteria must be *adequate* and valid, i.e., must assess what a researcher wants to assess.
3   The criteria must be *neutral* with respect to studied phenomena. For instance, assume that during a pedagogical experiment schoolchildren in a class study a new topic, while the ones in the other do not. Then the level of knowledge of this topic could be the criterion for their comparison.
4   The whole set of the criteria must cover all essential characteristics of a studied phenomenon or process with sufficient *completeness*.

One may also face a somewhat different (yet, legitimate) interpretation of the notion of a criterion. Notably, a criterion is understood as the qualitative side of the obtained result or achieved goal. Accordingly, the notion of a criterion is separated from that of an indicator or parameter. In such interpretation the same criterion may have several indicators or parameters. For instance, the efficiency (a criterion) of fulfilling a certain task by a worker or specialist is estimated by means of time spent and errors made (parameters).

The step of choosing assessment criteria for validity of research results finalizes the conceptual stage of research design. The next stage of research lies in modeling (hypothesis construction).

THE STAGE OF HYPOTHESIS CONSTRUCTION. Hypotheses construction represents a primary method of scientific knowledge development. The method consists in generating a hypothesis with its subsequent experimental (or even theoretical) checking. As the result, the hypothesis is confirmed (thus, it becomes a concept or theory) or rejected (hence, a new hypothesis should be put forward). A hypothesis *per se* is a model of future scientific knowledge (possible scientific knowledge).

A scientific *hypothesis* acts in two roles, *viz.*, as a supposition regarding a certain form of the relationship between observed phenomena and processes or as a supposition regarding the relationship between observed phenomena, processes and their internal base. The first kind is called *descriptive hypotheses*, while the second one is referred to as *explanatory hypotheses*. As a scientific supposition, a hypothesis differs from arbitrary conjecture in that it satisfies some requirements. Meeting these requirements forms the justifiability conditions of a hypothesis.

The first *justifiability condition* of a hypothesis. A hypothesis must explain the whole range of phenomena and processes it is constructed for (i.e., for the whole problem domain of a corresponding theory). Moreover, a hypothesis should not contradict the previous facts and scientific statements (as far as possible). However, imagine that one fails to explain given phenomena based on existing facts; then it is necessary to hypothesize by contradiction to earlier statements.

The second justifiability condition of a hypothesis is the fundamental *verifiability of a hypothesis*. A hypothesis is a supposition regarding a certain indirectly observed foundation of phenomena. It can be verified by comparing the deduced corollaries with the existing experience. That a hypothesis is unavailable for experimental checking means its unverifiability.

The third justifiability condition of a hypothesis lies in *applicability of a hypothesis* to a wider possible range of phenomena. Given a hypothesis, one must be able to deduce those phenomena and processes described by this hypothesis. Furthermore, it should serve for deducing a wider possible class of phenomena and processes (being not directly related to the initial ones).

The fourth justifiability condition of a hypothesis is the maximal possible fundamental *simplicity of a hypothesis*. This has nothing in common with easiness or primality. The actual simplicity of a hypothesis makes up its ability (in terms of a common *basis*) to explain a wider possible range of various phenomena or processes without artificial constructions and arbitrary assumptions (new hypotheses put forward in each specific case).

Meeting the four justifiability conditions of a hypothesis does not mean that the latter becomes a theory. Yet, otherwise a supposition may not claim to be a scientific hypothesis.

In addition to the above justifiability conditions of a hypothesis, we should emphasize a series of important aspects. In particular, a hypothesis must be formulated only within the problem domain which includes the problem posed by a researcher. For instance, in many Doctoral theses (in social sciences, as well as in technical and natural sciences) one observes shifts of problem domains during hypotheses construction. As a result, a thesis becomes indistinct and vague; a researcher has a poor understanding about what he/she actually does.

Any hypothesis can be fruitful only if (until investigations are finished) a researcher uses it similarly to the knowledge being generally accepted in science. In other words, this takes place if a researcher proceeds from a hypothesis as an established system of knowledge. Otherwise, a scientist appears unable to reason in a rigorous and consistent way, to make concrete logical deductions and check them empirically. Otherwise, he/she would fail to identify the points where inferences (hypotheses deductions) vary with the established facts and impede the search of new facts.

An investigator must be prepared for generating new hypotheses, as well as for choosing and analyzing alternative hypotheses. Indeed, quite often science provides an explanation to the same phenomena and processes based on different hypotheses. Reviewing such hypotheses requires much time and energy to solve complex problems (empirical, theoretical, logical ones). The presence of alternative hypotheses is an important prerequisite of science development, since it allows avoiding preconception in interpretation and usage of obtained results.

The following stage of research design (research planning) proceeds from a defined goal of research, chosen assessment criteria and a constructed hypothesis. It comprises the steps of decomposing (determining the tasks of research), the step of analyzing the conditions (available resources), and the step of making up the program of research.

THE STAGE OF RESEARCH PLANNING. *The step of decomposing (determining the tasks of research).* As is generally known, a *task* is a given goal of a certain activity in specific conditions. Thus, the tasks of research act as partial (relatively independent) goals of research in concrete conditions used to check a constructed hypothesis. As a rule, the tasks of research are formulated in two ways.

The first way (actually, the simplest and nonrigorous one being feasible, e.g., to write Candidate's theses) consists in the following. Tasks are stated as rather separate and complete stages of research. Nevertheless, they have few things in common with scientific problems proper; most likely, tasks represent procedural components of research. Many verbs such as "to study," "to analyze" assist in formulating tasks. And so, we explicitly obtain the stage-type temporal structure of research tasks planning. Notably, each subsequent task admits solution only based on solution of the previous one.

The second way appears more sophisticated and rigorous in scientific sense (and so, it is preferable). Similarly, the tasks of research are stated as relatively independent and complete stages of research. But in contrast to the first way, it seems difficult to identify the temporal sequence here. Tasks are dictated by the necessity of solving separate subproblems (with respect to a research problem) and achieving partial goals, i.e., subgoals (with respect to the main goal of research). Of course, tasks are formulated in concrete conditions within a generated hypothesis of research.

The step of analyzing the conditions (available resources). Any feasible scientific problem can be solved merely under definite conditions (in a special case, under available resources). For a complete list of *conditions of activity* with a detailed description, the reader is referred to [29]. For instance, we have to mention personnel-related, motivational, material and technical, methodical, financial, organizational, regulatory and legal, and informational conditions.

No doubt, one should carry out a deep analysis for each task of research and each group of above conditions. In particular, it is necessary to answer the following questions. What concrete conditions exist for solving a specific task? What conditions must be satisfied or provided? Performing research activity, an investigator should focus on personnel-related, material and technical, and informational conditions.

The step of making up the program of research. The final step in the stage of research planning lies in making up the program (technique) of research. *The program of research* is a document describing all relevant components or research (a problem, an object and subject, a goal, a hypothesis, tasks, methodological principles, and methods). Moreover, making up the program of research implies planning, i.e., research work scheduling. Many scientists take a skeptical view of research planning; nevertheless, the existing experience indicates that planning is useful for organization and self-organization.

Discussing the aspects of planning, one should have in mind two types plans; *viz.,* the matter concerns individual research planning and collective research planning.

Individual planning. Note that research planning requires certain skills being developed with the lapse of time. A young investigator possesses no experience, and he/she needs a proficient scientific advisor. For the sake of fun, an author of this book (A.N.) recollects the following episode. As an associate research officer, he submitted a draft 1-year program of research to his scientific advisor. Who would have thought then that implementing the plan takes 23 years and yields Dr. Sci. degree!

The issues of collective research planning will be studied in Chapter 5 (they are inseparably linked with organization of such research).

THE STAGE OF TECHNOLOGICAL PREPARATIONS FOR RESEARCH consists in making up experimental documentation (textbooks, learning aids, observation report sheets, questionnaires). In addition, this stage covers purchasing or production of necessary experimental assemblies, the development of necessary software, and so on. Technological preparations are specific for each scientific research.

Therefore, we have discussed all stages and steps of the design phase of a scientific research. Now, let us consider the technological phase of research.

4.2 TECHNOLOGY OF SCIENTIFIC RESEARCH

The technological phase of research lies in direct checking of a constructed scientific hypothesis. Technological phase comprises two stages, i.e., the stage of research implementation and the stage of results summarization.

THE STAGE OF RESEARCH IMPLEMENTATION. It includes two steps known as the theoretical step (analyzing and systematizing the publications, perfecting the conceptual framework, defining the logical structure of the theoretical part of research) and the empirical step (carrying out experiments and tests).

Theoretical step. Analyzing and systematizing the publications. A permanent review of scientific literature represents an obligatory component of any scientific activity. Moreover, scientific literature is an important tool for the existence and development of science proper. First, it serves for distributing and storing the obtained scientific knowledge. Second, scientific literature acts as a means of communication among researchers. One has to account for different functions of certain types of publications (they reflect various development stages of a scientific knowledge).

In Section 2.1 we have underlined that new scientific facts, ideas or theories first appear in the form of abstracts at scientific conferences, seminars, congresses, and symposia. Alternatively, they are published as preprints or other materials issued rapidly. Being subsequently reviewed and systematized, new scientific results form scientific papers in periodicals (journals, lecture notes and collections of articles). The next stage of their generalization, systematization and verification leads to *monographs*. Finally, the most important (fundamental and general) components of a scientific knowledge are published as textbooks for students and schoolchildren. The stated dynamics of a scientific knowledge must be considered by a researcher during publications review; it is necessary to identify different sources by their relevance, authenticity and recognition in a scientific community.

Beginning a literature review, an investigator compiles a *bibliography*.

Each research requires defining the primary scientific concepts or theories as its basis. The matter concerns not all scientific publications cited in a research work (they can be tens or hundreds). On the contrary, one should choose (at most) four concepts proposed by famous scientists, underlying the research.

Furthermore, an investigator must have a clear image of the methodological foundations of his/her research. The above need is also connected with the following feature. Generally speaking, science includes different *scientific schools*, often treating the same problems from various positions. Scientific schools may have nonidentical (or even opposite) opinions. Still, the existence of diverse scientific schools is objectively necessary for science development. Nevertheless, constructing his/her research, an investigator must demonstrate a tough posture on the basic and supplementary theories and concepts used in the research.

A major requirement applied to any research is involving a rigorous, precise and univocal *terminology*. In everyday life, we operate terms freely; science calls for an ordered and rigorous language.

When a researcher has to use a certain term, he/she should study a corresponding entry in general dictionaries, encyclopedias and thesauri. The mentioned sources provide a unique interpretation for terms being in general use at a nation-wide level. As a rule, different dictionaries suggest almost identical interpretations for a term; yet, each dictionary introduces some subtle nuance in explanations (thus, facilitating its usage). Note that a researcher may take advantage of Webster's Dictionary or Collins Dictionary to find synonyms or substitutes for certain terms (in order to avoid tautology and enrich the language).

The next step is refining purely philosophical, epistemological and methodological notions (by addressing proper dictionaries).

In philosophical dictionaries any researcher would benefit by getting acquainted with key notions (categories) such as abstraction, an analysis, a knowledge, a value, quality, quantity, a model, an observation, a norm, an explanation, generalization, an image, an object, an experience, a base, a relation, practice, a subject, a problem, development, reflexion, semantics, a system, systems analysis, a property, comparison, essence, similarity, a theory, a form, formalism, an experiment, and others.

It might be useful to know the terminology of *logic*: abstracting, an abstraction, an axiom (the axiomatic method), an algorithm, an analogy, a correlation, ascending from the abstract to the concrete, a hypothesis, epistemology, deduction, a law, a sign, a knowledge, an idea, invariance, induction, information, a study, a class, a classification, composition, a component, a context, a concept, a tuple, logic, the logical and the historical, a measure, a meta-theory, a direct knowledge, consistency, notion generalization, the law of inverse relation, a common notion, the volume of a notion, the definition of a notion, the special, a relation, an estimate, a parameter, a notion, a postulate, the rules of introducing a notion, synthesis, an attribute, a principle, a problem, a contradiction, a procedure, the content of a notion, comparison, a structure, a term, a type, a condition, a fact, etc.

And the final step is analyzing the interpretation of terms (being relevant for a concrete research) in scientific literature. To begin, one should study fundamental publications of the authors whose theories or concepts underly the research (recall the above discussion). It appears reasonable to compile a corresponding *thesaurus*. A thesaurus is a dictionary of selected terms used by authors, revealing their interpretation and mutual correlation. Later on, while writing reports, papers, books or theses, a researcher involves terminology mostly from this thesaurus. The rest of the terms are engaged only as the need arises (if managing without them is impossible). Still, by applying a certain term, an investigator performs self-control by adhering to a definite interpretation (for auxiliary terms) and substantiating the choice of a specific interpretation (for primary terms).

The "danger" of introducing new terms waylays any researcher. Sometimes this is very attracting. However, scientists are unwilling to accept new terms in science (or even have a guarded look on them). This is natural – a *language* (including a scientific language) represents national heritage requiring a proper care. Suppose that every researcher adopts new terminology in his/her publications; this would lead to total misunderstanding among scientists (and people, as well). Thus, introducing new terms is admissible only if absolutely necessary (when none of the existing terms describes a corresponding phenomenon or process). It is utterly prohibited to assign a new meaning (an "author's definition") to generally accepted terminology.

Let us again discuss a *conceptual framework*; an important aspect should be mentioned here as follows. The choice and systematization of the conceptual framework in a concrete research both depend on its subject, as well as on the formulated goals and tasks. Therefore, the essence of phenomena and processes being expressed via a fixed system of concepts is defined by author's position. In addition, an author predetermines the conceptual system for his/her research (another "pair of shoes" is that this system might be precise and rigorous or indistinct and contradictory).

Constructing the logical structure of a theoretical research. With the exception of the process of constructing the *logical structure* of a new scientific concept or theory (see a detailed discussion below), building the logical structure of a theoretical research or the theoretical part of an empirical research appears variable and completely depends on the subject, goal and tasks of a specific research. Merely some aspects are common; we consider them in the current section.

To construct the logical structure of a research, one often has to adopt different *classifications* and introduce new classifications. Moreover, classifications are desirable as they make a research well-composed. Here are the major **requirements applied to a classification:**[2]

1   A classification can be made only using a single basis. Probably, this is the principal requirement (yet, often ignored). Introducing a certain classification, a researcher must explicitly specify a corresponding basis. A *basis for classification* is an *attribute* enabling to split the volume of a *generic concept* (the whole set of classified objects) into types (*specific concepts*, i.e., the elements of this set). For instance, the basis used to divide schools into (a) elementary schools (primary schools) (b) junior high schools and (c) senior high schools secondary schools (high schools) lies in the level of general education demonstrated by a student of a corresponding school. At the same time, within a certain classification it seems impossible to divide school students by their age and results (or by the results and attendance of optional classes).

---

[2]A well-known example of a lame classification (violating the stated requirements) is called The Animal Classification by the Chinese court sages. "Animals are divided into a) the ones belonging to the Emperor b) embalmed ones c) tamed ones d) suckling pigs e) sirens f) fairy-tale ones g) stray dogs h) the ones included in this classification i) the ones behaving violently as being mad j) innumerable ones k) the ones drawn in a very thin brush from camel's wool l) and the rest ones m) the ones that have just broken a jug and n) the ones resembling flies from afar."

2   The volume of classification elements must coincide with the volume of the whole class being classified. Suppose that we have partitioned all triangles into acute-, right- and obtuse-angled ones (the basis for classification is triangle angles). There exist no other elements (types of triangles) in this classification.

3   Each object enters only a single subclass. For instance, it is not allowed to classify all integer numbers into even, odd and prime numbers. Otherwise, numbers $5, 7, 11, \ldots$, would simultaneously enter two subclasses (as being odd and prime numbers).

4   Classification elements must be mutually exclusive; i.e., none of them is included in the volume of another. For example, scientific books could not be divided into monographs, textbooks, handbooks and books on mathematical issues. Indeed, the latter may be monographs, textbooks or handbooks.

5   A classification must be continuous – one should choose the closest subclass ("jumping" to an outlying subclass is prohibited). Scientific investigations can be classified as the ones in the fields of physics, chemistry, biology, ecology, etc. By no means could scientific investigations be classified in the same way as the ones in the fields of chemistry, biology, ecology and electrodynamics (a branch of physics). Here we have "jumped" from the closest subclass (physics) to the distant one (a branch of physics).

Furthermore, the same class of objects, phenomena or processes can have different classifications with respect to various bases. For instance, furniture may be classified depending on

–   basic material (wooden, metal, or plastic furniture, etc.);
–   design style (classical style, the Empire style, Victorian style, art nouveau, and so on).
–   color (black, white, brown furniture, etc.);
–   functionality (tables, chairs, cabinets, . . .).

Thus, the same object may have different bases for classification.

Building the logical structure of a research work, an investigator inevitably faces "logical crossroads" and has to choose a certain direction for further development. There might be many such crossroads, and trying all possible roads is naturally impossible (even a whole life can be insufficient). And so, a researcher chooses a unique road to-be-considered as the primary one (promising good prospects). Assume that the "crossroad" is of crucial importance for the whole research work; accordingly, an investigator substantiates his/her choice. Anyway, one should not excuse for having not performed something. Indeed, scientists possessing rich experience in the field of research logic construction surely know about such "logical crossroads"; a well justified choice is always natural.

It happens that a researcher has to divide his/her logical constructions with respect to different classifications (in the view of various aspects). Such a situation baffles everybody. Indeed, how can relevant aspects be described without repeats? This is impossible! Thus, a certain aspect or classification should become the basic one, and the other aspects are discussed within the framework of the basic aspect. Unfortunately, this approach appreciably reduces the richness of exposition; but there is no alternative.

Table 4.2 An example of the system of classifications for interaction of a new chemical compound.

		Types of compounds	
		Acids	Alkalis
Temperatures	Low	+	+
	High	+	
	Normal	+	+

Finally, note that a set of classifications with respect to different bases (special bases exist to identify them) is said to be a *system of classifications*. The issues of building and analyzing systems of classifications play an important role in the logical structure of a theoretical research. Notably, they allow explicitly outlining a corresponding *problem domain* (it defines the basis for classification of the bases for the system of classifications, see Fig. 4.4). In addition, they allow identifying interconnected subdomains (in a given problem domain) and choosing "blank spots" (promising subjects or methods of research). Finally, studying all classes of a certain basis assists in generalization – see Fig. 4.5.

Consider an elementary example. Suppose that the subject of research consists in properties of a new chemical compound. The bases for classification include (1) the types of compounds interacting with a given one (here we separate out acids and alkalis) and (2) interaction temperatures (low, high and normal temperatures). The problem domain (the properties of the new compound when interacting with other compounds) comprises six interconnected subdomains – see Table 4.2. These subdomains intersect as the value of a corresponding attribute varies.

Assume that during the research an investigator has examined the cases marked by the symbol "+" in Table 4.2. Consequently, the research turns out incomplete (the case of new compound's interaction with alkalis under high temperatures has not been analyzed). Imagine that the new compound has been discovered to interact with acids identically (under low, high or normal temperatures). Since just three values of the attribute "temperature" have been identified, one can generalize as follows. The interaction of the new compound with acids does not depend on temperature.

A system of classifications can be modified by eliminating current bases for classification and/or adding new ones. In the above example, it is possible to add the basis for classification called "the concentration of compounds interacting with a given compound." Thus, the subject of research becomes wider.

Therefore, systems of classifications represent an efficient logical tool ensuring the wholeness and completeness of the subject of a research. They are useful for an investigator (in organization of scientific activity, it is necessary to order a problem domain, to understand what has been done and what has to be done). Such logical tools are attractive as a form of representing the results of research (e.g., favouring the rapid comprehension of the derived results by a reader).

Building the logical structure of a theory (a concept). First, let us distinguish between the notion of a *theory* in a certain science and the notion of a *scientific theory*. The former is interpreted as the whole set of theoretical knowledge in a given field

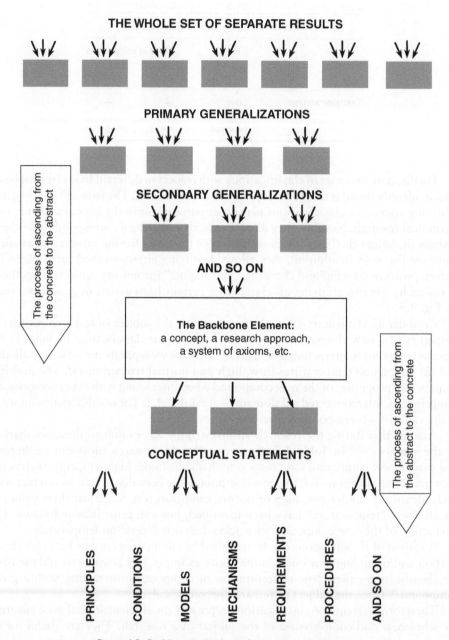

Figure 4.5 Building up the logical structure of a theory.

of science (e.g., physics, biology, etc.). At the same time, each scientific field includes numerous scientific theories (concepts). As a matter of fact, almost any high-quality Doctor's thesis provides an integral theory or concept. Here we discuss the development of scientific theories (concepts).

The process of constructing the logical structure of a theory or concept consists of two stages. The first stage lies in *induction*, i.e., ascending from the concrete to the abstract, when a researcher has to determine the backbone element in his/her theory. This can be a concept, a system of axioms or axiomatic requirements, an integral research approach, etc.

We should emphasize the following. The term "concept" possesses two meanings. First, it acts as a principal *idea*, a message of something. Second, a concept serves a synonym of a theory. Here, we adopt this term in both interpretations. Notably, the first meaning is to consider a concept as a short (yet, rich in content) formulation. The second meaning takes place during deployment and development of a concept (in its short formulation) in the set of conceptual statements, principles, factors, conditions, mechanisms, and so on.

At the inductive stage, epistemologically weak sciences have the classification approach as the only means of generalization. A researcher seeks for corresponding *bases for classification* allowing to unite, combine and generalize the existing results.

During the process of results' generalization, an investigator has to address his/her problem domain in the aspect of theory completeness. What "hollow spaces" are observed in the problem domain? They must be filled (probably, via additional experimental work or borrowing the results obtained by other scientists). On the other hand, an investigator has to correlate the derived results and problem domain with the set of theoretical (again, in the aspect of completeness and consistency of a theory or concept).

At the inductive stage, a researcher traces out all available results being of crucial importance. Next, using definite bases for classification, he/she combines them in primary generalizations. Afterwards, in a similar way a researcher gets secondary generalizations, and so on. The described inductive process is known as *abstracting* (ascending from the concrete to the abstract). It takes place until all results are reduced to an author's *concept* (a short meaningful statement of 5–7 rows) reflecting the whole set of results, the essence of a research work. Note that a system of axioms or an integral research approach can be used instead of a concept, as well.

Suppose that the stage of induction (determining and formulating the backbone element – a concept, a research approach, a system of axioms, etc.) is concluded. It passes on the baton to a deductive process referred to as *concretizing* (ascending from the abstract to the concrete). This stage is remarkable for the following. The formulation of a concept gets developed and deployed in the set of principles, factors, conditions (groups of conditions), models, mechanisms, etc. Sometimes the problem of research is decomposed into several relatively independent aspects; accordingly, a concept regenerates as a series of conceptual statements. Subsequently, the latter evolve in the form of sets of principles, and so on. On the other part, principles can advance into classes of models, types of problems, etc. This is how the logical structure of a theory is built, see Fig. 4.5. An investigator often addresses the presented diagram cyclically to check and refine the logic of his/her research.

Unfortunately, in rich literature on epistemological issues we have not found an hierarchical system of structural elements of a theory. For instance, which element has a higher (lower) level of hierarchy in the sense of abstracting (concretizing) – a principle or model, a rule or requirement, a mechanism or procedure? Evidently, an investigator may assign his/her own *hierarchy* according to specific tasks of research (traditions of a scientific organization he/she belongs to).

Therefore, a *theory* (a *concept*) is the backbone element representing (in its narrow sense) a system of axioms, etc., including its conceptual statements and other concretizing structures known as *structural elements* of a theory.

Let us enumerate the **structural elements of a theory** (this can be useful for researchers): an algorithm; a tool (didactic tools, conceptual tools, etc.); classifications; criteria; methodologies; methods; mechanisms (classes of mechanisms); models (basic models, prediction models, graph models, open models, closed models, dynamic models, complexes of models, etc.); directions; substantiations; bases; basics; paradigms; parameters; periodizations; approaches; notions (developing notions, systems of notions, etc.); techniques; principles; programs; procedures; solutions; systems (hierarchical systems, generalized systems, etc.); a content; ways; means; schemes; structures; strategies; essences; taxonomies; tendencies; technologies; typologies; requirements; conditions; phases; factors (backbone factors, etc.); forms (sets of forms, etc.); functions; characteristics (essential characteristics, etc.); goals (sets of goals, hierarchies of goals); stages, and so on.

Moreover, epistemologically strong sciences also include theorems, lemmas, and assertions. A backbone element may be a theory, a concept, an idea, an integral research approach, a system of axioms or a system of axiomatic requirements, etc. In some sciences (e.g., chemistry, pharmaceutics, microbiology), a backbone element may consist in the fact of synthesizing a new chemical compound, a new medication, a new vaccine, etc. (as the fruit of many years' work of a researcher). The conditions and principles of their application are discovered afterwards.

Generally speaking, the logical structure of theories or concepts appears common.

Empirical step. Experimental work. A specific feature of a scientific research lies in the following. *Experimental work* serves merely for confirming or rejecting preliminary theoretical constructions (hypotheses); yet, it occupies considerable time available to an investigator.

One would think that the experimental part of a research starts as soon as an investigator has identified and derived all theoretical constructions. Nevertheless, generally a reseacher joins in experimental work much earlier. Indeed, prior to organizing and conducting experimental work proper (exactly those experiments confirming or rejecting a hypothesis), an investigator should master the basic skills of planning and performing experimental work, analyzing and generalizing its results. Moreover, such a preliminary stage allows choosing necessary approaches, fine-tuning corresponding tools, etc.

We have already mentioned that experimental work is specific for each concrete research; indeed, it totally depends on the content of a concrete research and would hardly be described in a general form. The only thing is that there exist rather universal tools to plan and process experimental results, known as *data analysis* and *statistical methods* (for instance, see [16, 17, 29, 31, 32] and other sources).

THE STAGE OF RESULTS SUMMARIZATION. A concluding stage of the technological phase of research includes results' approval, formulation and publication.

The step of approving the results. A detailed *approbation* of a research represents a condition of its justifiability and validity of corresponding results. Furthermore, this is a real way to correct and eliminate research shortcomings. The word "approbation" has Latin origin and *expressis verbis* means "approvalor praise." Here critics, opponents or judges make up scientific colleagues, practicians, and whole research groups. Approbation takes place in the form of public reports, presentations, discussions, as well as in the form of written or oral reviews. Note that informal approbation (conversations and disputes with colleagues, experts in other fields of scientific knowledge and practicians) is also important. Based on the results of approbation, a researcher comprehends and considers the existing issues, positive and negative estimates, objections and advice. And so, he/she refines research materials or revises some statements of the research (if necessary).

The step of formulating the results. Approbation being successfully completed, a researcher formulates and publishes the results of research. Actually, a *publication* (a written, oral or electronic one) is a prerequisite to conclude a scientific research (of course, if the latter is really scientific). Notably, a new knowledge obtained by a researcher constitutes a scientific knowledge only by being public heritage.

The results of a conducted research are published in the following forms.

1   An *essay* is an initial form of written presentation for research results. Using an essay, young scientists express the original results of their investigations. As a rule, an essay describes the theoretical and practical relevance of a topic, analyzes corresponding publications, as well as includes assessments and conclusions regarding the studied scientific sources. An essay must demonstrate a sufficient level of erudition of a researcher, his/her skills to analyze, systematize, classify and generalize the existing scientific information. Generally, essays are not published.

2   A *scientific paper* represents the most common and wide-spread form of publications. Scientific papers appear in periodicals (scientific journals and collections of articles). The exposition of a scientific paper must be systematic and consistent. Different sections must be logically connected. A particular attention should be paid to the scientific style of a work. In fact, scientific style has the following primary requirements: clear exposition, accurate word usage, laconicism, rigorous scientific terminology, successive presentation of key positions, consistency, interconnected statements. Of crucial importance are the issues of text editing.

In addition, one should be careful when formulating scientific conclusions and suggestions. The corresponding part of a paper serves for (a) brief and precise identification of essential aspects in research results and (b) demonstration of their possible implementation in practice.

3   A scientific *report*. A research work can be published as a scientific report.

The following major requirements apply to a scientific report: precise structuring; the consistency (logic) of material presentation; cogent arguments; compact and accurate formulations; concrete statements of research results; demonstrable conclusions and well-grounded recommendations.

As a rule, appendices include auxiliary material of a report (numeric tables; examples of instructions, manuals, questionnaires, tests that have been developed and used in a research work; supplementary illustrations, and others).

A *scientific presentation* has much in common with a scientific report. Meanwhile, in contrast to the latter, the former may cover only a certain logically completed part or aspect of a research.

4   A *textbook*. A textbook proceeds from theoretical *recommendations* for perfecting a certain process (e.g., teaching and educational process, a technological process, etc.); note that such recommendations are based on the research results. Since a textbook is intended for practicians, it should have a good (living) literary language. Wherever possible, a textbook should include visual materials.

   A textbook appears in the form of a brochure or book. A *brochure* represents a small printed matter (5–48 pages) with a softcover or without a cover. A *book* is a nonserial printed matter having over 48 pages; generally, a book includes a cover or binder.

   A classical example of a brilliant textbook on military art is *The Science of Win* by A. Suvorov. Just imagine – a warlord of genius used merely 25 pages of text to provide numerous valuable recommendations (from rules of warfare and military marches to rear services and hospitals' organization).

5   A *monograph*. A monograph is a scientific publication focusing on a certain problem and treating it in a comprehensive and integral way. A monograph may have one or several authors.

   Suppose that a researcher has managed to solve a certain problem in a new fashion, to generalize the existing scientific works on this problem. Moreover, assume that he/she can substantiate the suggested concepts on the problem and demonstrate concrete opportunities of their practical implementation. In this case, a researcher should publish his/her results as a monograph.

   In fact, a monograph describes how a studied problem was treated earlier (in scientific literature and practice) and how it is solved nowadays. Next, a monograph reveals the essence of the author's approach to the problem, states the methods of research used by him/her to justify the proposed concept. Subsequently, a monograph elucidates and analyzes the results of author's investigations, followed by reasonable conclusions and science-based recommendations. Finally, a monograph provides the corresponding bibliography. A monograph is published in the form of a brochure or book.

6   Abstracts of reports presented at conferences, seminars, etc. Generally, during scientific conferences or seminars, the collections of report abstracts are printed. Abstracts represent a very small document (1–3 pages of text). The maximum volume of abstracts is determined by the organizing committee of a conference or seminar. Here the primary task is explaining the basic results of a research in a compact recapitulating form (the results would be expanded during the author's report at a conference, seminar or symposium).

In addition to written forms of research results' presentation, there also exists oral communication among scientists. Let us give definitions to the following forms of oral communication:

–   a scientific (problem-oriented) *seminar* is a small-group discussion of scientific reports and messages prepared by its participants under the supervision of a leading scientist or expert. There are regular and one-time scientific seminars. They appear to be an important factor of cohesion in a research group; moreover, seminars assist in designing common approaches and views in such groups. Scientific seminars take place in a scientific organization or educational institution (the representatives of other organizations may be invited though). Classical examples of regular seminars are famous *Pavlov's Wednesdays* (the materials were published as multivolumed sets), as well as Landau's Seminar on theoretical physics;

– a scientific *conference* is a meeting of scientists and/or practicians in a certain field. Scientific and practical conferences are always topical; they take place at the level of a scientific organization or educational institution, at the regional, national or international level;

– a scientific *congress* is a meeting of representatives of a whole scientific branch at the national or international level (e.g., a congress of psychologists). Most modern problems being relevant to a given science are considered there;

– a *symposium* (Greek *sumposion*, *sumpinein* which means "to drink together") is an international meeting of scientists and experts on a narrow issue (problem);

– an *authoring school* of the best experience (workshops, training sessions, etc.) is a form of communication among scientists and practicians, when an author of the best (scientific or practical) experience shares it with school participants. Authoring schools are conducted within a certain organization, enterprise, educational institution (alternatively, at the regional or national level).

Therefore, we have presented the sequence of steps in the technological phase of research (from a research idea to results' summarization and publication).

4.3   REFLEXION IN SCIENTIFIC RESEARCH

Prior to discussing the reflexive phase of scientific research, let us introduce the notions of an estimate and reflexion.

An *estimate* is a relation to a phenomenon, activity or behavior, establishing their relevance or correspondence to some norms and goals; assessing the degree, level or quality of something. Note the term "estimate" designates both the process and result of such assessment.

The term *"reflexion"* was first suggested by J. Locke. However, in different philosophical systems (the ones by J. Locke, G. Leibniz, D. Hume, G. Hegel, etc.), reflexion possessed various interpretations. Reflexion (from Latin *reflex* which means "bent back") is:

• a principle of human thinking, guiding humans towards comprehension and perception of one's own forms and premises;

• subject consideration of a knowledge, critical analysis of its content and cognition methods;

• the activity of self-actualization, revealing the internal structure and specifics of spiritual world of a human.

There are three types of reflexion, namely,

– elementary reflexion, leading to consideration and analysis of a knowledge and actions, to thinking about their limits and value;

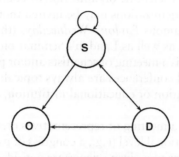

Figure 4.6 The ways of estimating.

– scientific reflexion, representing the criticism and analysis of a theoretical knowledge based on methods and techniques of a given field of scientific knowledge;
– philosophical reflexion, being the comprehension and perception of limiting bases of objective reality and thinking, human culture in the whole.

Generally, philosophical literature understands reflexion as addressing cognition to itself, as thinking about thinking. To put it in simpler words, reflexion has the following aphoristic definition: "Reflexion is an idea of an idea." In this book, we *par excellence* touch the issues of elementary reflexion.

The reflexion of a subject, i.e., his/her thoughts about his/her own thoughts of reality, activity, etc., is said to be *self-reflexion* or *reflexion of the first kind*. We emphasize that most social research works concentrate on self-reflexion.

Reflexion of the second kind takes place with respect to other subjects (includes thoughts of a subject about possible thoughts of another subject).

Roughly speaking, the complicated process of reflexion has (at least) six positions characterizing the mutual representation of subjects. They are: (1) the subject existing in reality, (2) the subject being seen by him/her, (3) the subject being seen by another subject, and (4)–(6) are similar to (1)–(3) (but in the view of another subject). Therefore, reflexion makes up the process of double mirror-type self-imaging by subjects of themselves and the reality. However, the number of such self-imagings can be large. To provide a common description, let us consider interrelations among three elements (see Fig. 4.6), *viz.*, the subject of activity (S), the object of activity (O) and different subjects (D). The arrows designate separate acts of "imaging."

We use various sequences of chars ("S," "O," and "D") to describe relations among the elements. The order of chars corresponds to (a) who assesses what or (b) who performs reflexion about what.

The relations of the first order (*zero-rank reflexion*) include the following estimates:

SO – the estimate of the results of subject's activity by himself/herself (*self-appraisal of the results*);
SS – the estimate of the subject by himself/herself (individual *self-appraisal*);
SD – the estimate of other subjects by the subject (as individuals);
DO – the estimate of the results of subject's activity by other subjects;
DS – the estimate of the subject by other subjects (as an individual).

Being passive, the object appears unable to estimate; moreover, we do not consider self-appraisal of other subjects (DD). Therefore, the above five relations exhaust feasible combinations of the relations of the first order.

These relations may be the subject of thoughts for the subject of activity and other subjects. This yields first-rank reflexion.

The relations of the second order (first-rank reflexion). Here one should distinguish between:

– self-reflexion (reflexion of the first kind), which corresponds to SS-type sequences, i.e., subject's thoughts about his/her self-appraisal and self-appraisal of his/her results:

> SSO – the subject's thoughts about his/her results' self-appraisal;
> SSS – the subject's thoughts about his/her self-appraisal.

and

– reflexion of the second kind (the rest sequences):

> SDO – the subject's thoughts about the estimate of his/her activity results by other subjects ("what others think about the results of my activity");
> SDS – the subject's thoughts about the estimate given to him/her by other subjects ("what others think about me");
> DSS – other subjects' thoughts about the subject's self-appraisal;
> DSO – other subjects' thoughts about the subject's self-appraisal of his/her activity results;
> DSD – other subjects' thoughts about the estimate given to them by the subject.

The relations of the third order (second-rank reflexion). Naturally, in this case we find numerous combinations. Some are provided below: SDSO – the subject's thoughts about other subjects' thoughts about his/her self-appraisal of the subject's results ("what others think about my estimates of my results"); DSDO – other subjects' thoughts about subject's thoughts about the estimate given to his/her activity results by other subjects, and so on.

In the framework considered, one may infinitely increase the order of relations (i.e., reflexion rank). However, the goals of this book allow involving (at most) second-rank reflexion. The application of more sophisticated reflexive processes (including the corresponding mathematical models and techniques, as well as illustrative examples) is discussed in [30].

Note that the processes of estimation and reflexion are inherent not only to the reflexive phase of a project (a scientific project, a practical project, an art project, or an educational project). During an activity, a subject permanently estimates the intermediate results and performs reflexion with respect to these results, the technology of his/her activity, the estimate given to his/her technology and results by other subjects, etc.

Let us revert to the *reflexive phase* of research. The main point lies in that an investigator (or a group of investigators) must apply reflexion to the obtained results.

It is necessary to "address back" for comprehending, comparing and estimating the initial and terminal states of:

- the object of activity – *self-appraisal of results*;
- the subject of activity (his/her own self-appraisal).

The estimates and appraisals given to current and final results of a research work by other scientists (colleagues, reviewers, and opponents) exert a considerable impact on the estimate and self-appraisal of the obtained results. For instance, by definition any (Candidate's or Doctor's) thesis is the individual work of a researcher; yet, it often accounts for opinions of many discussants (a scientific advisor, research officers of a laboratory, lecturers from a department). Thus, a thesis is the fruit of collective work.

Moreover, self-appraisal of research results strongly depends on results' recognition by a corresponding scientific community or practicians. For this, one has to publish the results.

However, *publications* vary. The easiest and thankless way consists in manuscript deposition. Unfortunately, few people read deposited manuscripts (although, the latter are considered as fully-fledged publications).

The "demand" for a publication mostly depends on the precise and convenient (simple) presentation of the corresponding material. As a rule, an author hardly guesses what form would be convenient for readers. Indeed, any experienced scientist knows this aspect. It happens that a serious research (the fruit of many years' work) is published, and scientific community provides a neutral feedback. On the other hand, writing a paper extempore ("serving out" the same results from another perspective, even unexpectedly for the author) may draw a wide response.

Public recognition of a performed research lies in the fact of successful defence of Candidate's or Doctor's thesis. In the sequel, the research work is assessed by its citation index; the latter shows how often other authors cite a given research. Actually, citation index seems somewhat formal. For instance, some publications are not available to a wide audience. This can be a purely theoretical work or a historical microstudy. Nevertheless, in many countries the prestige of scientists (including the salary) directly depends on their *citation index*.

An important part in popularization and public recognition of research results is assigned to different forms of oral communication among scientists (e.g., participation in conferences, seminars, etc.). We emphasize that the forms of written and oral communication among scientists (publications and conferences, respectively) should run in parallel. Experience indicates that oral messages at conferences or symposia enable attracting the attention of a scientific community to existing research results. Moreover, such messages increase the interest in corresponding publications. The oral form of scientific communication is remarkable for the following, as well. Each conference includes a program with plenary and regular sessions; however, a researcher mostly benefits by informal communication with colleagues (during coffee breaks, banquets, stay in a hotel, etc.). According to sociological estimates, formal communication at a conference (reports, presentations, etc.) yields merely 30% information, whereas the remaining 70% result from informal communication.

In addition to assessment of research results by a scientific community, self-appraisal and reflexion of one's own research results is valuable.

We have already mentioned that self-appraisal and reflexion run through the entire activity of an investigator (from an idea to results' publication). This is the specifics of scientific activity.

Yet, *self-appraisal* and *reflexion* with respect to a completed research are of paramout importance for scientists. Put several questions. What are the positive and negative features of my research and why? Why the obtained results appreciably vary from the initial idea (such situation is common)? What theoretical constructions are superfluous (missed)? Are the methods of empirical research used correctly (sufficiently)? What issues should not have been treated (to save time)? And so on.

The answers should be taken into account in future work. A real scientist initiates a new project immediately after completing the current one (the cycle repeats). Accumulating the personal scientific (methodological) experience based on each finished research leads to "spiral-type" research development.

The above aspects of reflexion in a research work concern the so-called "elementary reflexion". We should also discuss **scientific reflexion**.

Scientific (or theoretical) reflexion over a system of scientific knowledge means its theoretical analysis, introducing a series of assumptions and idealizations, modeling of studied phenomena and processes. Scientific reflexion yields a new system of knowledge, being a relatively true reflection of actual relationships and implying several assumptions (arising at the modeling stage). Reflexion over the previous system of knowledge causes transcending it and generating a new knowledge. For instance, theoretical reflexion assisted G. Galileo in subjecting to criticism Aristotelian premises (assumptions) on the system of views of things. Relativity theory by A. Einstein revealed many hidden premises of classical Newtonian mechanics (they were even unclear to I. Newton and his followers). As a matter of fact, scientific reflexion is the interconnection between a previous knowledge and a new knowledge, between an "old" scientific theory and a "new" one. The continuity of a scientific knowledge is exactly the content embedded in the comprehension of the principle of correspondence – a fundamental principle of a scientific knowledge (see Section 2.1). The basic method of scientific reflexion is *retrospective analysis.*

The reflexive phase concludes a research work as a cycle of scientific activity (as a scientific project).

Up to this point, we have mostly considered individual research. The issues of organizing and implementing collective research possess some specifics; they are analyzed in the next chapter.

Chapter 5

Organization of collective scientific research

This book considers methodology as the theory of organization of an activity; here the emphasis is made on an activity performed by a subject (an individual or a collective one) and its completed cycle – a project. At the same time, an activity may have a sophisticated (internal subjective) structure. The corresponding examples are the problems of organizing a collective research (discussed below), the problems of organizing (controlling) a joint activity, and others. A detailed study of specific features arising in a joint activity goes beyond the present book. Actually, this is a promising direction of further methodological investigations.

Scientific literature provides many publications on the issues of collective research. However, they are mostly dedicated to managerial, bibliometric, psychological and sociological aspects. Here we are concerned with organization of *collective scientific activity*. Unfortunately, we have not found numerous published sources on this topic. Therefore, the material presented in Chapter 5 represents our own experience in the supervision of research groups (including large groups).

Naturally, to organize a collective research, it is necessary to choose a "director" of research. The following requirements are generally applied to a *director of research* (a *scientific adviser*):

1   First and foremost, he/she must master the methodology of scientific research, possess first-hand experience of investigations and have a definite authority in a scientific community.
2   On a purely voluntary basis, he/she must form a group of researchers and train them in the methodology of scientific research and explain the details of certain research.
3   He/she must plan the whole complex of scientific investigations being necessary at a given stage (including interaction with scientific foundations, other research groups and industry). Moreover, he/she must organize and assist in planning of individual research for each member of the research group, exercise control of plan fulfillment. In addition, he/she must generalize the obtained results. Thus, a director of research should be not only a good scientist, but also a talented manager!
4   Finally, he/she must plan and organize the publication and application of the derived scientific results, as well as plan next-step research.

Supervisors of a research group set themselves the following question. How can a common topic of collective research be formulated? The definition of a common topic for the whole collective makes up a significant psychological complicacy. On the one hand, working on a common topic allows uniting a research group and obtaining considerable (weighty) results.

On the other hand, any creative researcher has one's own range of scientific interests (which not necessarily agrees with a common topic). Thus, a director of research should have the craftsmanship to convince other investigators in the necessity of involving them in a common work. A director of research must possess sufficient skills and breadth of scientific views to see and find feasibilities of combining the interests of a separate researcher and common interests of a research group. According to our experience, a director of research succeeds in this by having a flexible, patient and persistent position. However, the major prerequisites are (a) all members of a collective work are carried away with research and (b) all members of a collective work have a clear image of the results desired by them and by the director of research.

An essential feature of collective scientific activity consists in different ability levels of its members. A director of research must account for this.

No doubt, abilities of people vary in any field of their activity. For instance, during educational process at schools different levels or qualities of certain teachers are "compensated" by classes' schedule. Yet, such "scheduling" is impossible in research organization. Furthermore, members of a research group would have nonidencial inclinations (e.g., a first member better manages investigations, another is a master of experiments; a certain member can be a good writer, while the other copes with oral presentations). Therefore, a director of research must carefully study individual attributes of different members (to take advantage of their skills and not to require impossible-doing tasks from them).

A director of research must adhere to the following important principle. Each member of a research group (except engineering staff, e.g., laboratory assistants) should have a separate area of research (i.e., an independent topic), be fully responsible for it and manage the corresponding results (including their publication under his/her name). Only in this case the members of a research group would work wholeheartedly.

Joint authorship (i.e., a paper or book is published under many names) is reasonable in science only as an exception, when the problem at hand allows a collective solution (and each author made a real contribution in its solution). It happens that a scientific adviser does not withstand the temptation of assigning his/her name to a research work performed by his/her subordinates. There is a well-known moral aspect of this phenomenon. Moreover, such a scientific adviser damages his/her scientific authority and prestige. Indeed, when the same co-author appears in many publications on totally dissimilar issues, a scientific community draws "appropriate conclusions" about such a scientist.

A director of research has his/her own area of scientific activity to show research skills (including publications) without infringing upon the interests of other members. A director of research works "at a different level"; notably, he/she formulates the common topic or hypothesis of collective research, generalizes the results of partial investigations, analyzes the observed tendencies and tasks for further research, etc. In fact, this is a large and independent (extremely interesting!) area of work.

In addition, a director of research may be in charge of a specific topic (thus, acting as an ordinary member of a research group).

Suppose that a director of research has determined a common topic of collective research. Subsequently, he/she prepares the general program of research in the form of a short text document. Here the general goals and directions of research are indicated.

As a rule, all topics of research works (within a collective research) must enter the general topic as certain components and become the elements of the research program.

Note that the *object*, *subject* and *goal* of a collective research are stated similarly to an individual research. The difference lies in its more general scale – the objects and subjects of separate research works represent some aspects or directions of a corresponding collective research. The goals of independent investigations can be treated as subgoals ensuring the achievement of a common goal of research.

A *hypothesis* of an individual research is rich in content and related to a certain problem. Contrariwise, a hypothesis of a collective research would be rather connected with assumptions regarding possible directions and aspects of the whole complex of forthcoming investigations. The problems of a collective research should be considered as the goals of individual research works.

The described preparatory operations being finished, a supervisor has to draw plans of collective scientific research.

In this context, we emphasize the following features of research planning.

1   Each topic starts with designing the methods of research.
2   Necessary activities are planned in maximum detail (including due dates); this enables discussing the obtained results at each stage, as well as controlling the progress of works. For instance, one must avoid situations when (on the expiry of 3–5 years) an executor claims, "I'm sorry, the hypothesis has been rejected, no results derived." Using yearly plans, a director of research should ask executors to present quarterly reports.
3   Plan necessary research activities so as each member of a research group understands his/her place and personal work. Again, avoid situations when a certain research activity (topic) has two or three coexecutors (actually, the work is performed by one member, whereas the rest "hide behind his/her back." Another inadmissible case is when somebody misappropriates the results obtained by others.
4   Interconnected work must be planned in the following way. Supervisors and executors of subsequent research activities (according to the logic of investigations) must be able to start based on intermediate results (not waiting for finalization of previous research results).

In addition, a plan includes the following sections:

–   scientific and organizational activities. This section serves to plan training hours for professional development of scientists (members of a research group), organization of scientific seminars and conferences. Moreover, this section includes planning of preparatory work to enter a post-graduate or postdoctoral course;
–   publishing activities. This section provides a list of works to-be-published (including deadlines);
–   activities to apply the derived results in practice.

A draft plan must be discussed in detail with all members of a research group. First, each member of the group must (inwardly and psychologically) accept it. Second, each member of a research group must understand his/her personal role and activities in the total amount of work. Third, discussing a draft plan, a research group must assess the feasibility of fulfilling the work in time.

After constructive discussion, a director of research approves perspective plan and annual plan. Next, it is necessary to develop and confirm individual plans of scientific activity for each member of a research group. An individual plan has an arbitrary form. The most important thing is that all activities (specified in the perspective plan and annual plan) are mentioned in individual plans. Finally, individual plans are signed by the executors and approved by the director of research.

Further activity of a director of research consists in control of plans' fulfillment and regular discussion of derived results. Naturally, for large-scale research projects, initial plans may be impracticable (during their realization, one discovers mistakes, faces new conditions, rejects some original hypotheses, etc.). The art of a real scientific adviser is to discuss in good time and introduce necessary corrections (in the plan, content and organization of research), to rearrange logical relations among separate directions of activities, and so on. Discussing the progress and intermediate results of investigations is important in the sense of generating common viewpoints, approaches and positions of researchers. Such discussions should take place at special scientific seminars.

Moreover, the head of a seminar must abide by the following rules of scientific disputes:

1   Each discussant has the right for a personal opinion, the right to express and persist in it. Any suppression of debates is strongly prohibited. Science does not solve problems by the majority of votes!
2   Only one discussant speaks simultaneously. He/she is not interrupted, having opportunity to express the opinion completely.
3   A speaker may be asked any questions concerning his/her activities. Only such formulations as "Am I correct, . . ." or "Please, elucidate . . ." are used.
4   Discussions touch the issues of what has been done by a speaker (not what might have been done by a discussant in the place of a speaker!). Any scientist has one's own view and comprehension of a problem. It is necessary to appreciate what has been done (instead of what is desired to-be-done by somebody).
5   A head of discussion guides debates according to their agenda (in a delicate, yet strict way). Digressing from the primary topic is not allowed. At the end of a discussion, the head must generalize and briefly present the results and outline further tasks.

In the course of research, its director is responsible for supplying additional resources. Indeed, implementation of plans always reveals the absence or lack of some resources. To solve a problem, members of a research group address their director. And he/she should have a proper attitude and settle all difficulties if necessary.

At subsequent stages of research, an important function of a supervisor of a research group lies in result generalization. For this, he/she regularly presents summary reviews and generalizing reports at seminars and meetings. It is reasonable to involve a scientific adviser as a scientific editor (to prepare publications, general scientific reports and presentations). First, a scientific adviser may observe the complete picture of obtained results. Second, by negotiating the editor's remarks with authors, one succeeds in integrating separate "parts" into the logical whole.

An invariable component of any collective research consists in the *expertise* of a completed research work. There exist internal and external expertises. The former is the public one conducted by the members of the research group. The latter takes place when a complete scientific report or program is submitted to a third-party scientific organization (an independent expert), a neighboring scientific institution or a higher educational establishment.

Finally, an essential direction of activities in a research group is implementing the obtained results (practical *application*). Experience shows that result publishing appears insufficient for their practical usage. Instead, one has to find a group of practicians being interested in a studied problem domain. Next, special "experimental platforms" are created at enterprises, firms or educational institutions to take advantage of the obtained results. Consequently, colleagues representing some neighboring enterprises or firms get to know about existing innovations. Where can we read about these innovations? Whom should we address for a consultation or advice? Thus, the implementation network gets mushroomed gradually. The stated aspect of results' application must be in the mind of any director of research. As a matter of fact, the ultimate goal of a scientific work lies in practical development. However, this concerns the so-called "practical sciences" (and has nothing to do with astronomy or mathematics).

Conclusion

Throughout this book, we have repeatedly compared different aspects of organizing research activity. In the Conclusion, we endeavor to compare them in a systematic way (in the logic of the key statements presented earlier). Notably, to succeed we analyze the basic characteristics, logical structure and organization of research process (its temporal structure) – see Tables C.1–C.3. Note that compiling summary tables represents a convenient analysis method. On the one hand, such tables appear useful for authors – numerous positions of a book are naturally refined. On the other hand, summary tables are helpful for readers as visual aids illustrating the primary content of a given book.

The authors would be grateful to readers for any constructive remarks and suggestions on the content of the book.

Table C.1 Characteristics of research activity.

Characteristics	Organization of research activity
Features of activity	1. Limited goals of a research work; a goal is formulated in advance; 2. Continuity of research; 3. Rigorous conceptual and terminological frameworks; 4. Mandatory publication of research results; 5. Pluralism of scientific viewpoints; 6. Communications in science; 7. Result implementation in practice (applications).
Principles of activity	Principles of scientific cognition: 1. The principle of determinism; 2. The principle of correspondence; 3. The principle of complementarity.
Conditions of activity	Motivational, personnel-related, material and technical, methodical, organizational, financial, regulatory and legal, and informational conditions.
Norms: 1. general; 2. specific	Universal ethical, hygienic and other norms. Norms of scientific ethics.

Table C.2 The logical structure of scientific activity.

Structural components	Organization of scientific activity
Active subject	A researcher.
The object of activity	The object of research decomposed by the researcher.
The subject of activity	The subject of research as an idealized aspect of the object, as a mental structure built by the researcher.
The result of activity	An objectively new scientific knowledge.
The forms of activity organization: 1. by the number of participants; 2. by the organization of activity process; 3. specific forms	Individual and collective forms. 1. Organizational culture as a universal form of activity organization. Modern design-technological type of organizational culture. 2. Projects as completed cycles of activity, their phases, stages and steps. Scientific schools as forms of organization of collective research activity. Institutional forms of organization of collective research activity: sectors, laboratories, departments, chairs, faculties research institutes, universities, etc.
The methods of activity: 1. theoretical methods-operations;	Mental operations: analysis, synthesis, comparison, abstracting, concretizing, generalization, formalization, induction, deduction, idealization, analogy, modeling, gedanken experiments, imagination.
2. theoretical methods-actions;	Dialectics (as a method), the method of knowledge systems analysis, scientific theories (as a method), proofs, the deductive (axiomatic) method, the inductive-deductive method of theories' construction, the identification and elimination of contradictions, problems statement, hypotheses formation, research approaches.
3. empirical methods-operations;	The analysis of publications, documents and results of activity, observations, measurements, (oral/written) inquiries, expert evaluation, testing.
4. empirical methods-actions	The methods of object tracking: investigations, monitoring, experience study and generalization; the methods of object transformation: trials, experiments; the methods of object analysis in the course of time: retrospection, prediction.
The means of activity	Material, informational, logical, mathematical, and linguistic means.

Table C.3 The forms of organization for the structure of scientific activity (*the life cyle of a research project as the temporal structure of activity*).

The forms of organization			
Phases	Stages	Steps	A research project
1. Design phase	1.1. Conceptual stage	1.1.1. Identifying contradictions	A scientific contradiction in practice or in a scientific knowledge system.
		1.1.2. Stating a problem	A scientific problem as the knowledge of a problem ("the lack of knowledge").
		1.1.3. Defining the goal of research	Specifying the goals of research. Generally, the goal is determinated by the problem and subject of research.

(continued)

Table C.3 Continued.

The forms of organization

Phases	Stages	Steps	A research project
		1.1.4. Choosing criteria	The criteria of scientific knowledge: 1. the general criteria of scientific knowledge: knowledge validity, intersubjectivity, systemacy; 2. the assessment criteria for validity of theoretical research results: single-subjectedness, completeness, consistency, interpretability, verifiability, validity; 3. the assessment criteria for validity of empirical research results are often defined by the researcher based on certain rules. The method of expert evaluation is also involved. Results validity is confirmed via statistical criteria.
	1.2. Modeling stage	1.2.1. Forming a hypothesis (constructing a model)	The cognitive model: a hypothesis as a conjectural scientific knowledge, as a model of possible new scientific knowledge (system of knowledge).
		1.2.2. Refining a hypothesis (optimization)	Concretizing a scientific hypothesis during research. Generally, a single hypothesis is verified.
	1.3. The stage of research planning	1.3.1. Decomposing (determining the tasks of research)	Formulating the tasks of research as the goals of solving separate subproblems according to a common goal of research and a stated hypothesis (the researcher has definite freedom of choice).
		1.3.2. Analyzing the conditions (available resources)	–
		1.3.3. Making up the program of research	Working out the program of research.
	1.4. The stage of technological preparations for research	1.4.1. Technological preparations	Making up experimental documentation (textbooks, learning aids, observation report sheets, questionnaires). Purchasing or production of necessary experimental assemblies, the development of necessary software, etc.
2. Technological phase	2.1. The stage of research implementation	2.1.1. Theoretical step	Theoretical stage of research: 1. analyzing and systematizing the publications; 2. perfecting the conceptual framework; 3. constructing the logical structure of theoretical research.
		2.1.2. Empirical step	Carrying out experiments and tests.

(continued)

Table C.3 Continued.

The forms of organization

Phases	Stages	Steps	A research project
	2.2. The stage of results summarization	2.2.1. Approving the results	Approbation in the form of public reports and presentations at conferences, seminars, symposia, etc.
		2.2.2. Formulating the results	Formulation and publication of research results as scientific printed matter (papers, monographs, textbooks, etc.).
3. Reflexive phase: Estimation (including self-appraisal and estimation of results) and reflexion			The critical analysis of research results; public recognition of research results; wide application of research results in practice. Reflexion as a way of comprehending the integrity of one's own activity, its goals, content, forms, methods, and means; as consecutive movement in the reflexive sense: "stop," "fixing," "abstracting," "objectification," and "reversion." Scientific reflexion as a way of building new knowledge systems.

Appendix. The role of science in modern society

Nowadays, one clearly observes a swift reappraisal of the role of science in human development. Let us endeavor to identify the causes of this phenomenon and discuss primary tendencies in further science development and interrelations in the traditional pair "science–practice."

First, we address some historical facts. As far back as in the Renaissance, science overshadowed religion to become the leading component in human ideology. Previously, solely hierarchs were able to judge about Weltanschauung; gradually, that role was entirely captured by the scientific community. Notably, the scientific community dictated its will and rules for almost in any field of public life; science was the supreme authority and truth criterion. For several centuries, scientific research appeared the basic activity cementing various professional fields of human beings. Science was the most important and essential institute; indeed, it defined the uniform image of the world and general theories. Partial theories and corresponding problem domains of professional human activity were identified with respect to the uniform concept of the universe. Scientific knowledge represented the "center" of society development; moreover, the generation of such knowledge was the basic type of production (actually, it predetermined the capabilities of other types of material and spiritual production).

However, the second part of the 20th century was remarkable for discovering cardinal contradictions in society development (both in science itself and social practice). We consider them below.

The contradictions in science.

1   The structural contradictions in the uniform image of the world created by science, internal contradictions in the structure of scientific knowledge (caused by science itself), the appearance of ideas regarding scientific paradigms shifts (the works by T. Kuhn, K. Popper and others).

2   The break-neck growth of scientific knowledge, accompanied by technologization of the corresponding tools of scientific knowledge production, resulted in segmentation of the image of the world. Accordingly, professional areas split up into numerous specialities.

3   Modern science is strongly differentiated; moreover, it turns out polycultural. In the past, all cultures were described within the common framework of European scientific tradition. Presently, each culture aspires to its own form of self-description and self-determination. The feasibility of providing the uniform

image of world history has become extremely problematical and condemned to mosaicity. The practical issues of organizing and controlling the "mosaic" society are immediate. It appears that traditional scientific models "operate" in an extremely narrow range. Notably, they are applicable in the fields connected with separating the uniform and general attributes (and not in the fields requiring the reflection of different things as indeed different ones).

4　And, above all, in recent decades the role of science (in the wide sense) has significantly changed with respect to social practice (also, in its wide sense). The triumph of science has gone. From the 18th century to the middle of the 20th century, there were many scientific discoveries; practice followed the pace of science by "picking them up" and implementing in social (material or spiritual) production. But that stage came abruptly to an end; as a matter of fact, the last epoch-making scientific discovery was the development of a laser (USSR, the 1950s). Gradually, science was "switching" to technological perfection of practice. The notion of scientific-technical revolution was replaced by that of technological revolution (technological epoch, etc.). Thus, scientists focused on perfection of technologies. For instance, consider rapid development of computer engineering and information technology. According to "general science," a modern PC has no fundamental differences against its first counterparts of the 1940s. At the same time, we observe appreciable reduction in its size, performance increase, and memory. Recall new languages of interaction between a PC and human beings, as well. The provided examples demonstrate that the focus of science shifts towards technologies (direct servicing of practice).

Formerly, theories and laws were in common use. Contrariwise, nowadays science rarely reaches such level of generalization. Most attention is paid to models being characterized by numerous possible solutions of problems.

Historically, there exist two major approaches to scientific research. The first one was suggested by G. Galileo. According to his viewpoint, science aims to establish an order underlying different phenomena (in order to represent the capabilities of objects generated by the order and to discover new phenomena). In fact, this is the so-called "pure science" (theoretical cognition).

The second approach was proposed by F. Bacon. It is not often thought of. However, exactly the corresponding viewpoint has prevailed recently: "I work for future well-being and strength of the mankind. To succeed, I offer science being efficient not in scholastic disputes, but in inventing new handicrafts ..." Modern science follows the path of technological perfection of practice.

5　From time immemorial, science generated "everlasting knowledge" used by the practice (i.e., laws, principles, or theories functioned for centuries or decades). But recently science has switched over to "situational" knowledge (especially, in public and technological fields).

This feature is mostly connected with the principle of complementarity (see Section 2.2). This principle appeared as the result of new discoveries in physics at the junction of the 19th and 20th centuries; during this period, it was found that a researcher studying an object introduces certain modifications in it (e.g., by a device used in the experiments). The principle of complementarity was first stated by N. Bohr: "Opposites are complementary." Notably, integrity reproduction for a phenomenon requires the application of mutually exclusive "complementary" classes of concepts during cognition. In physics this means that acquiring the experimental data about certain physical quantities is invariably connected with modifying the data about other quantities being complementary to the former. Complementarity serves for establishing the equivalence between the classes of concepts providing a complex description to contradictory situations in different fields of cognition.

The principle of complementarity considerably altered the system of science. Classical science operated as an integral system intended for (a) obtaining the set of knowledge in the final and completed form, (b) eliminating from the scientific context the impact of researcher activity and the means used by him/her, and (c) assessing the absolute validity of the knowledge included in the science fund. This situation was changed by the principle of complementarity. Here we acknowledge the following important aspects. Embracing the subjective activity of a researcher by the scientific context modified the essence of knowledge subject. Instead of the "pure" reality, the subject of knowledge became a certain "section" of the reality defined in the light of accepted theoretical and empirical means and ways of reality cognition by a subject. Moreover, the interaction between a studied object and a researcher (e.g., using devices) definitely leads to different levels of displaying the object's properties depending on the type of interaction with the cognizing subject (in different, often mutually exclusive conditions). This implies the legitimacy and equivalence of different scientific descriptions of the object (various theories concentrated on the same object or problem domain).

Second, many modern investigations take place in applied domains (e.g., economics, engineering, education, etc.). They are devoted to designing optimal situational models of organizing industrial, financial, educational structures, firms, and so on (note that optimality is considered under given specific conditions). Yet, the results of such research are actual for only a short time. Once the corresponding conditions vary the above models are no more necessary. Anyway, this science is useful, and such investigations are fully scientific.

6   Next, we have earlier employed the term "knowledge" designating scientific knowledge. Presently, one adopts different types of knowledge (in addition to scientific knowledge). For instance, the ability of managing a text editor represents a complicated knowledge. But it would hardly be a scientific knowledge (just imagine the appearance of a new text editor – this knowledge will be relegated to oblivion). Other examples include databanks and databases, standards, statistical indicators, train or schedules, huge information collections in Internet, etc. (in fact, we use such knowledge in everyday life). In other words, today a scientific knowledge coexists with other (unscientific) knowledge.

Contradictions in practice. The development of science (in the first place, natural science and engineering knowledge) ensured industrial revolution to the mankind. As the result, by the 1950s people have almost solved the primary problem of their history (the problem of starvation). For the first time in its history, the mankind succeeded in subsisting and creating a favorable way of life (we mean the majority of people!). And so, the mankind passed to a totally new (the so-called post-industrial) era of development. The latter is remarkable for abundance of food products, commodities, and services (leading to an intense competition in the world economy). Therefore, significant deformations happened around the world (in politics, economics, a society or culture, etc.). An inevitable attribute of the new era was instability or dynamism of political, economical, public, legal, and technological situations. Everything in the world was subjected to continuous and swift changes. Hence, practice must adapt to new conditions. The innovation of practice becomes a time attribute.

Several decades ago, in the conditions of a relatively stable way of life, social practice and practicians (engineers, agronomists, physicians, teachers) could easily wait for science to develop new recommendations and approve them via experiments (the next stage was waiting for product designers and engineers to create and approve the corresponding structures and technologies). When all was said and done, the matter concerned practical application of the results. Today, such waiting turns out to be pointless. Indeed, a situation changes dramatically during this period of waiting. Thus, the practice (naturally and objectively) selected an alternative way; practicians started suggesting innovative models of social, economic, technological and educational systems themselves (author's models of production processes, firms, organizations, schools, technologies, methods, etc.).

In addition to theories, intelligent entities such as projects and programs were revealed in the previous century. Furthermore, by the end of the 20th century the activity regarding their creation and implementation has become wide-spread. They are supported by analytical work rather than by theoretical knowledge. Using its theoretical strength, science itself generated the ways of mass production of new sign forms (models, algorithms, databases, etc.); that was the stuff for new technologies of material and sign production. Generally speaking, technologies (along with projects, programs) became the leading form of activity organization. The specifics of modern technologies lie in that none of theories or professions is able to cover the whole technological cycle of a certain production process. Complex organization of large-scale technologies results in that former professions correspond to just one or two stages of large technological cycles. To make a career, a man must represent a professional being able to join in these cycles (actively and competently).

Yet, for skillful organization of projects, the development and application of new technologies or innovative models, practicians require the scientific style of thinking; the latter comprises many essential qualities such as being dialectical, systematic, analytical, logical and broad-minded regarding problems and feasible consequences of their solution. Most importantly, they require the skills of research work, in the first place – the skills of rapid orientation in informational flows and construction of new models. The matter concerns either cognitive models (scientific hypotheses) or pragmatic, i.e., practical, innovative models of new systems (economic, industrial, technological, educational systems and others[1]). Probably, this is the general reason of aspiring for scientific research by all practicians (managers, financial analysts, engineers, teachers, etc.) as a worldwide tendency.

Thus, the aforesaid implies that in modern conditions science and practice draw closer.

---

[1]Indeed, modern industrial technologies are changed in 5–7 years. Naturally, it seems impossible to predict them and train the corresponding personnel. Thus, any specialist should comprehend new information rapidly and be broad-minded (for the complete list of necessary qualities, see the above discussion). Such qualities are developed only by involving in research activity.

In organization of both scientific and practical activity (first of all, productive and innovative activity), one would easily observe many common features. Notably, they are constructed using the logic of projects. A project proceeds from an idea resulting in a model as a certain image of future system (a new system of scientific knowledge in the case of a research project; a new industrial, technological, financial or educational system in the case of a pragmatic or practical project). Next, the model is analyzed and (possibly) implemented. Historically, project-based organization of an activity originated in the Renaissance (at that times, art was separated from handicrafts, and creation of an art work was remarkable for project features). Of course, the notion of a project and project-based organization of an activity has appeared recently. Evidently, in scientific investigations project-type organization of an activity has finally gained its place at the junction of the 19th and 20th centuries. Notably, a mandatory attribute of most research works was the presence of a hypothesis (i.e., a cognitive model); accordingly, a research work became a project. In contrast, project-based organization of practical activity consolidated its positions in the second part of the 20th century.

Meanwhile, organization of scientific activity (on the one part) and organization of practical activity (on the other part) possess essential differences; a fundamental one lies in the following. In a research work, it seems impossible to uniquely define the goal of a specific project. A new scientific knowledge must appear only as the result of such activity (implementation of a project). There is a precise formulation of initial material (i.e., scientific knowledge accumulated by the moment a research work starts). Hence, a certain paradox arises: to organize an activity (a research project), one should have a terminal goal as a normative result of activity (a result of project implementation). However, it is impossible to define the goal of a research work normatively. This goal is often posed in an inconcrete way (using general-purpose verbs such as "to study," "to define," "to formulate," etc.).

Similarly, in practical activity there is no concrete (definite) conception regarding the result of activity (the result of implementing a certain project). Nevertheless, the requirements towards the result (at least) approach the latter to a level of definiteness allowing to judge about implementability and innovation of a project. This level can be compared with that of similar projects (in their type and scale) or with real state of a certain process.

Actually, in modern conditions of societal development, science and practice resemble opposite sexes needed for human reproduction (further development of our civilization). Perhaps, science acts as female gender (being a subtle and capricious object) and practice acts as male gender (being a rough and straightforward object).

In science, the knowledge about the lack of knowledge seems to be (at least) of the same relevance as a positive knowledge. True, such results are often characterized by disproval. Even physicians are used to saying, "A negative result is still a result," mostly to console an unlucky colleague (passing over the negative result itself). Generally, in science the complexity due to misunderstanding is treated as an *ad interim* phenomenon being admissible. And a researcher can "maneuver" by changing the subject or method of research.

In practical activity, the complexity due to misunderstanding is often treated as an unacceptable thing causing inadmissible delays of problem solution. As a rule, practicians have to apply "a frontal attack" to a problem. This is why managers in any practical activity make intuitive and strong-willed decisions (that fail frequently).

Perhaps, the negative experience of such solutions results in the following. The way of thinking of managers and other practicians approaches that of scientists; the role of scientific methods in practical activity increases.

Apparently, the process of mutual approximation and "convergence" of science and practice is a characteristic feature of the present times.

Now, let us imagine what possible consequences of this phenomenon would be. Discuss the consequences for social practice and science.

The development of scientific potential of social practice and the growth of professional staff form a positive trend to be supported. Serious negative consequences (both for material and spiritual production) are still not noticeable. The matter seems much complicated with science and scientific community.

The consequences for science. By willingly assisting practicians in their scientific growth (sometimes, for mercenary motives), researchers "prepare a pitfall for themselves."

First, hundreds and thousands of theses are defended based on author's models of firms, financial structures, industrial processes, educational institutions – the corresponding results require theoretical interpretation, generalization, systematization, etc. (in order to join existing economical, pedagogical, mathematical and other theories). Unfortunately, scientists have not got down to this work. Yet, the amount of information increases.

Second, under the conditions of opinion pluralism, many scientists have been carried away by designing new directions in science (generally, these are "pseudo-new" directions, i.e., pre-existing principles are considered using some new values). For instance, in pedagogical science one observes numerous new "pedagogies" such as "anthropocentric pedagogics," "vitagenic pedagogics," "gender pedagogics," and so on (innovative pedagogics including "the pedagogics of love"). Of course, we must not totally reject the necessity of such research. But this "waters down" the body of scientific research; science grows "in bushes" (and not "in trunk").

Third, the stated factor is aggravated by the following. Recent years (again, due to the increasing number of theses) have been remarkable for rapid development of scientific potential of universities, special research institutes and academies of advanced training. This tendency is definitely positive. Moreover, we clearly observe the increasing amount of research accompanied with an expanding range of research directions. But poor scientific communication (low funding of business trips, small circulation of scientific journals, irregular scientific conferences and seminars) and coordination of scientific works make the field of feasible research (in many branches of scientific knowledge) almost invisible – and so, one would hardly go there.

Fourth, the dramatically increasing number of scientific investigations "waters down" the bounds among scientific schools. Previously, the relatively small amount of research and limited number of scientific schools enabled relating a new research work to a specific scientific school. Today, any new Doctor of Science (or even a Candidate of Science!) often searches for and selects followers (undergraduate and post-graduate students) to create a new "scientific school." Subsequently, these followers receive degrees and create their own "scientific schools." Thus, the process gets expanded. Moreover, in addition to the growth of "science immensity," short-term preparation of scientific personnel enhances scientific and methodological incompetence of new researchers. That is, most Candidate's and Doctor's theses are fulfilled and defended in short periods; thus, a potential scientist has no time to "grow" into the corresponding scientific environment (community) and to "absorb" the methodological culture of a research work. Having rapidly defended his/her thesis, a newly-fledged Doctor or Candidate then starts "training" new undergraduate and post-graduate students. Therefore, we obtain the game known as Chinese whispers or the game of telephone.

Fifth, there exist the so-called regionalization (self-isolation of scientific schools from world science) and scientific sectarianism. Being afraid of competition for certain resources (e.g., recognition, funding, etc.), a scientific school or a group of researchers hardly accepts the advances of other investigators.

Sixth, a curious paradox arises as follows. In the past, scientists and practicians were "at opposite (interconnected) poles"; notably, the first pole was occupied by "theory," whereas the other one belonged to "practice." Practicians stood agape to heed the voice of science. Nowadays, the situation is changing fast. Most practicians defend their theses and continue their practical work. Thus, a new "tandem" appears, with a professional scientist (at the one pole) and a practician combining his/her practical activity with scientific research. For convenience, we will call the former by a "scientist-theoretician," while the latter by a "scientist-practician." And they would talk "on equal terms." In such conditions, "scientists-theoreticians" may preserve their status (and the status of science) by raising their level of scientific generalizations (their theoretical level). However, most professional scientists would be hardly able to succeed in this. And so, the approximation of science and practice generates new serious challenges exactly for science (for the whole scientific community). How will these challenges be treated? Time will show!

Therefore, we summarize the ideas by stating that the role of science in modern society has changed dramatically. And this factor still exerts (and will definitely exert) a significant impact on all sides of life (politics, economics, social sphere and culture).

But an interesting paradox concerns exactly the process of education! We have already mentioned the following. Presently, the unstable conditions of social life (leading to the necessity of performing research works for almost any specialist – even in purely pragmatic fields) require scientific training. And the issue of such training (since one's schooldays) is immediate. Indeed, modern literature provides numerous publications regarding the involvement of schoolchildren in research activities (educational scientific projects). Scientific societies of students appear in colleges (although, the mission of these educational establishments is not preparing future scientists). Many universities provide courses on scientific research and related issues (intended for scientific and methodological preparation of investigators). Furthermore, yearly projects and diploma theses (degree projects) of students in colleges acquire the attributes of research works. Thus, the described process is wide spread in practical education. This direction can be referred to as *scientific education* (as a component or line of educational content). The emphasis shifts from training in a ready-made scientific knowledge towards mastering the techniques used to obtain the knowledge in question; in fact, the emphasis shifts towards the methodology of scientific research.

Bibliography

[1]   A Guide to the Project Management Body of Knowledge. 4th ed. (2008) Project Management Institute.

[2]   Ashby W. (1956) An Introduction to Cybernetics. London. Chapman and Hall.

[3]   Bell J. (1999) Doing Your Research Project (3rd ed.). Buckingham. Open University Press.

[4]   Bertalanffy L. (1968) General System Theory: Foundations, Development, Applications. New York. George Braziller.

[5]   Bird A. (1998) Philosophy of Science. New York. UCL Press.

[6]   Creswell J. (2002) Research Design. Qualitative, Quantitative and Mixed Method Approaches. London. Sage Publications.

[7]   Davis G. and Parker C. (1979) Writing the Doctoral Dissertation. A Systematic Approach. New York. Woodbury. Barrons Educational Series.

[8]   Dawson C. (2002) Practical Research Methods. New Delhi. UBS Publishers' Distributors.

[9]   Deal T. and Kennedy A. (1982) Corporate Cultures: The Rites and Rituals of Corporate Life. Harmondsworth. Penguin Books.

[10] Delbert M. and Salkind N. (2002) Handbook of Research Design and Social Measurement (6th ed.). London. Sage Publications.

[11] Jones J. (1970) Design Methods. New York. John Wiley & Sons.

[12] Goddard W. and Melville S. (2004) Research Methodology. An Introduction. Lansedowne. Juta and Company Ltd.

[13] Goode H. and Machol R. (1957) System Engineering: An Introduction to the Design of Large-scale Systems. Boston. McGraw-Hill.

[14] Handy C. (1985) Understanding Organizations (3rd ed.). Harmondsworth. Penguin Books.

[15] Hillier F. and Lieberman G. (2005) Introduction to Operations Research (8th ed.). Boston. McGraw-Hill.

[16] Kelly A. and Lesh R. (2000) Handbook of Research Design in Mathematics and Science Education. New Jersey. Lawrence Erlbaum Associates.

[17] Kothari C.R. (1985) Research Methodology Methods and Techniques. New Delhi. Wiley Eastern Limited.

[18] Krantz D., Luce R., Suppes P. and Tversky A. (1971) Foundations of measurement. Vol. I–III. New York. Academic Press.

[19] Kumar R. (2005) Research Methodology – A Step-by-Step Guide for Beginners (2nd ed.). London. Sage Publications.

[20]   Kuhn T. (1962) The Structure of Scientific Revolutions. Chicago. University of Chicago Press.

[21]   Lakatos I. (1976) Proofs and Refutations. Cambridge. Cambridge University Press.

[22]   Landau L. and Lifshitz E. (1976) Mechanics. Vol. 1 (3rd ed.). Oxford. Butterworth-Heinemann.

[23]   Leedy P. and Ormrod J. (2005) Practical Research. Planning and Design (8th ed.). New Jersey. Pearson Educational International and Prentice Hall.

[24]   Leont'ev A. (1978) Activity, Consciousness and Personality. Prentice. Prentice-Hall.

[25]   Lock D. (2000) Project Management. New York. John Wiley & Sons.

[26]   Marshall C. and Rossman G. (1999) Designing Qualitative Research. 3rd ed. London. Sage Publications.

[27]   Merton R. (1942) The Normative Structure of Science. In: R.K. Merton, The Sociology of Science: Theoretical and Empirical Investigations. Chicago, IL: University of Chicago Press, 1973.

[28]   Mesarovich M., Mako D. and Takahara Y. (1970) Theory of Hierarchical Multilevel Systems. New York. Academic.

[29]   Novikov A. and Novikov D. (2007) Methodology. Moscow. Sinteg.

[30]   Novikov D. (2013) Theory of Control in Organizations. New York. Nova Science Publishers.

[31]   Osborne J. (2008) Best Practices in Quantitative Methods. London. Sage Publications.

[32]   Panneerselvam R. (2004) Research Methodology. New Delhi. PHI Learning Pvt. Ltd.

[33]   Pfanzagl J. (1968) Theory of Measurement. New York. Wiley.

[34]   Phillips E. and Pugh D. (1987) How to Get a Ph. D.: A Handbook for Students and Their Supervisors (2nd ed.). Buckingham. Open University Press.

[35]   Poincaré H. (1908) The Foundations of Science. New York. Science Press.

[36]   Potter S. (ed.) (2002) Doing Postgraduate Research. London. Sage Publications.

[37]   Popper K. (1959) The Logic of Scientific Discovery. London. Routledge.

[38]   Rosenberg A. (2000) A Philosophy of Science: a Contemporary Introduction. London. Routledge.

[39]   Rubinstein S. (1981) Probleme Der Allgemeinen Psychologie. New York. Springer-Verlag.

[40]   Sage A. (1992) Systems Engineering. New York. Wiley IEEE.

[41]   Salmon M. (1992) Introduction to the Philosophy of Science. New York. Prentice Hall.

[42]   Shein E. (1992) Organizational Culture and Leadership: A Dynamic View. San Francisco. Jossey-Bass.

[43]   Taha H. (2011) Operations Research: An Introduction (9th ed.). Prentice Hall.

[44]   The Cambridge Handbook of Expertise and Expert Performance (ed. by K. Ericsson). (2006) Cambridge. Cambridge University Press.

Name index

Subject index

Communications in Cybernetics, Systems Science and Engineering

Book Series Editor: Jeffrey 'Yi-Lin' Forrest

ISSN: 2164-9693

Publisher: CRC Press/Balkema, Taylor & Francis Group

1. A Systemic Perspective on Cognition and Mathematics
    Jeffrey Yi-Lin Forrest
    ISBN: 978-1-138-00016-2 (Hb)

2. Control of Fluid-Containing Rotating Rigid Bodies
    Anatoly A. Gurchenkov, Mikhail V. Nosov & Vladimir I. Tsurkov
    ISBN: 978-1-138-00021-6 (Hb)

3. Research Methodology: From Philosophy of Science to Research Design
    Alexander M. Novikov & Dmitry A. Novikov
    ISBN: 978-1-138-00030-8 (Hb)

T - #0240 - 071024 - C0 - 246/174/8 - PB - 9780367380120 - Gloss Lamination